THE FRONTIERS COLLECTION

The books in this collection are devoted to challenging and open problems at the forefront of modern science and scholarship, including related philosophical debates. In contrast to typical research monographs, however, they strive to present their topics in a manner accessible also to scientifically literate non-specialists wishing to gain insight into the deeper implications and fascinating questions involved. Taken as a whole, the series reflects the need for a fundamental and interdisciplinary approach to modern science and research. Furthermore, it is intended to encourage active academics in all fields to ponder over important and perhaps controversial issues beyond their own speciality. Extending from quantum physics and relativity to entropy, consciousness, language and complex systems—the Frontiers Collection will inspire readers to push back the frontiers of their own knowledge.

More information about this series at http://www.springer.com/series/5342

Len Pismen

Morphogenesis Deconstructed

An Integrated View of the Generation of Forms

 Springer

Len Pismen
Department of Chemical Engineering
Technion
Haifa, Israel

ISSN 1612-3018   ISSN 2197-6619   (electronic)
The Frontiers Collection
ISBN 978-3-030-36816-6   ISBN 978-3-030-36814-2   (eBook)
https://doi.org/10.1007/978-3-030-36814-2

This Springer imprint is published by the registered company Springer Nature Switzerland AG
The registered company address is: Gewerbestrasse 11, 6330 Cham, Switzerland

Preface

This book is about morphogenesis as *genesis of forms*, not necessarily in a plant growing from a seed or an animal from an embryo (although they supply the most abundant and unequivocal material) but also in kindred processes, from inorganic to social. It is about our morphogenetic universe: unplanned, unfair, and frustratingly complicated but benevolent in allowing us to emerge, survive, and inquire into its laws. Why did I embark upon writing this book? Forms of life and related forms of inorganic Nature that allowed life to originate and evolve are, on the one hand, captivating philosophically charged themes and, on the other hand, a subject of intensive investigation into their minute detail. The first has prompted many fascinating books, some turning into bestsellers; the second, mountains of scientific papers and monographs, read by a few specialists if at all. My intention here is to interpose bit-size samples of the detailed view into the broad picture, in the attempt to raise the former and anchor the latter.

I have tried wherever possible to emphasize structural connections between dissimilar phenomena. One could find a deeper and more eloquent description of any particular subject in this book but I hope this approach will appeal to the inquisitive reader with a wide range of interests or perhaps to the wide-eyed youth who has not yet decided what to do in life. Although the content is not technical, the reader should be tolerant to unfamiliar scientific terminology, which I have tried to decipher and to make more friendly. It is impossible to avoid biological nomenclature with its unpronounceable Graeco-Latin terms, often of an obscure or unrelated origin, which seems to be designed to repel a lay reader, but you will not see here arbitrary names of genes, enzymes, and signaling molecules. In order to keep the chronology right, I often refer to original publications, even though later interpretations, derivative works, and reviews might be more accessible. The citation list cannot be comprehensive in such a wide area, but references may give the interested reader a thread to a plethora of related literature in the citation network.

Why the word *deconstruction* in the title? Google defines *deconstruction* as a method of critical analysis of philosophical and literary language. Deconstruction implies an outside viewpoint, and this is how the physicist views the intricacies of life, trying to infer general features from a confusion of details. In the words of Heidegger (1962), *Destruktion* aims to "arrive at those primordial experiences in which we achieved our first ways of determining the nature of Being". This attitude is not bereft of dangers. Bare etymology implies that deconstruction is apt to destroy what has been carefully constructed, in the way the meaning of a text used to be deconstructed by Jacques Derrida and his followers. In contrast, deconstruction is understood here in an affirmative sense more fitting of Heidegger's and Google's definitions as the analytic way to understand *construction*.

Len Pismen   *Haifa, 2019*

Contents

Chapter 1
Morphogenetic Universe

1.1 Unplanned, Unfair but Benevolent

The book of Genesis (King James version) starts succinctly: *In the beginning God created the heaven and the earth. And the earth was without form, and void; and darkness was upon the face of the deep.* Modern science views the birth of our Universe much in the same way, perhaps substituting spacetime for heaven and matter for earth but not offering much more detail. Indeed, what detail is needed when matter is without form, in a disordered state of thermodynamic equilibrium that prevailed till the Universe cooled down to become transparent, leaving us to see the cosmic background radiation cooled down to 3 degrees kelvin, as a remainder of the day when *God said, Let there be light: and there was light. And God saw the light, that it was good: and God divided the light from the darkness.*

Max Tegmark (2014) imagines that our Universe grew right after the Big Bang just as a baby human grows right after conception: "each of your cells doubled roughly daily, causing your total number of cells to increase day by day as 1, 2, 4, 8, 16, etc." – and in the same way unimaginably tiny Planck-size spacetime cells multiplied each unimaginably tiny time interval during the course of cosmic inflation. A baby's cell is, however, infinitely more complex. It carries an elaborate development plan in its DNA that predicts in many ways the baby's destiny. Ardent Bible readers may believe in a benevolent Creator who likewise immersed in the baby Universe a thoroughly planned morphogenetic program, to be realized during the ensuing days of Creation and beyond – but this cannot be reconciled with our scientific outlook.

The Universe at large seems to be haphazard. We can, of course, observe with satisfaction a nice symmetry in the Standard Model with its quarks, leptons, and gauge fields neatly fitting into their allotted places, and be thankful for lucky combinations of fundamental constants allowing our world to last for billions of years, thereby leaving enough time for the creation, first, of a long list of elements in the Periodic Table, and then, of a variety of their combinations in complex chemistry and elaborate crystalline structures. But in what a roundabout way it has been achieved! Little else but gravity was employed to gather together dispersed matter into dense clumps,

© Springer Nature Switzerland AG 2020
L. Pismen, *Morphogenesis Deconstructed*, The Frontiers Collection,
https://doi.org/10.1007/978-3-030-36814-2_1

some of which in due time heated up sufficiently to ignite nuclear fusion. The largest clumps, overzealous in this activity, exploded, sacrificing themselves to create the heavy elements indispensable for complex chemistry on cooler chunks of matter. Some of these had the good fortune to be placed in a fairly constant orbit around a moderately burning thermonuclear energy source, and eventually, as a miracle we still do not understand, carbon, hydrogen, oxygen, and nitrogen atoms managed to combine under benevolent stars into self-replicating structures that started to evolve toward ever increasing complexity.

Neither good luck nor thorough planning are required to create variety and complexity, one just needs to wait. The first source of complexity is *symmetry breaking* in its numerous forms: it may act to eliminate variety before recreating it on another level. Symmetry was broken in the very beginning when what we call *matter* was chosen against what we call *antimatter* – of course, what has been vanquished is always labeled anti-something. Further on, just two kinds of quarks, up and down, remained to dominate matter, with all others dispersed in occasional outbursts that engender cosmic rays or created at great expense in elaborate constructions built by highly evolved animals. Of all leptons, only electrons remained in neatly structured atomic shells and molecular orbitals. Of all gauge fields, only light was pronounced good on the First day of Creation and left to illuminate us ever since.

But already on the Second day the uniformity of matter was countered by divisions. *And God said, Let the waters under the heaven be gathered together unto one place, and let the dry land appear: and it was so.* This was a prototype of phase transitions and the formation of interfaces separating what cannot be mixed. Redistribution of means and creation of inequality was already anticipated by elimination of competitors to the standard components of matter, and presaging its social implementation, by the evangelical Matthew's law: *For whosoever hath, to him shall be given, and he shall have more abundance: but whosoever hath not, from him shall be taken away even that he hath.* Galaxies and stars were formed, as masses gravitated to other masses to create high density contrast. In accordance with the same aggregative tendency, large droplets grow at the expense of smaller ones which evaporate or dissolve; successfully adapting species displace others sharing the same niche; stronger bulls gather larger harems to perpetuate their genes; bestsellers leave less lucky books out of print and shows unattended.

Symmetry breaking and aggregation are powerful but crude tools. They served well to create the amazing forms of the inorganic world, and they still operate on higher levels as tools for pattern formation, most prominently in areas of scarcity, creating vegetation patterns in arid landscapes, segregated neighborhoods in troubled cities, and extreme wealth inequalities in poor countries. But the emergence of sophisticated forms in the later days of Creation required more refined instruments. *And God said, Let the earth bring forth grass, the herb yielding seed, and the fruit tree yielding fruit after his kind, whose seed is in itself, upon the earth: and it was so.* The key word here is *seed*. A seed implies *memory*; it carries a program transmitted from earlier generations. A living body does not vanish in vain but yields *fruit after his kind, whose seed is in itself.* This is cooperation extending in time, and we can add that cooperation, both in time and space, extends through life, to enzymes unraveling genetic code, to signaling among cells, to language and society.

1.2 Construction and Deconstruction

Construction – making things – and deconstruction – understanding how things, small and large, up to our world in its entirety, are made – are closely related, and have been developing together. Chimpanzees are able to use tools, but they just take them ready-made, similar to modern consumers buying ready-made gadgets, though rather more sophisticated than a stick or a stone. Neither do they engage in understanding the structure of these tools or trying to improve them; a chimp is aware of the properties of wood no more than we of the innards of our smartphones. Primitive humans go a step further. They hone a stick to a sharp arrow or spear; they populate the woods by spirits and fairies – not exactly a scientific endeavor but a sign of striving to acquire know-how.

The next stage, attained concurrently with agricultural evolution, is producing tools, and, in parallel, consolidating pantheons and contriving creation myths. The development of knowledge and proficiency became highly non-uniform. In any society that has advanced beyond a primitive level, knowledge is specialized, and what counts is knowledge and aptitude acquired by the society as a whole. Advanced populations split into settled cultivators and nomad herders, worshipping different gods, accustomed to different lifestyles, and deconstructing their world in different ways. Focus for a moment on Greece, fifth–fourth century BC, setting the stage for two millennia ahead, sculpting matchless images, erecting porched temples admired to this day, and trying to penetrate the way forms are constructed.

One way is to assemble them from distinct parts, just as ships crossing the Mediterranean are pieced together from boards, sails, and metal joints. This leads directly to the atomism of Leucippus and Democritus, imagining the world being built up of more abstract distinct entities, indivisible atoms of different shapes, moving in the void or packed into various arrangements. Another way is to cast a desired form by deforming a continuum, as craftsmen do by forging metal, or blowing glass, or molding clay, or (jumping ahead in time) 3D printing. This was the outlook of Aristotelian natural philosophy, which has been dominant longer than any non-religious belief.

Modern science and technology adopt both ways. Wares are assembled from parts, but eventually parts should be made from a raw material. Science, striving to discern internal structures, alternates between atomism and the calculus of the continuum. Discretizing reality in the Democritean tradition, organizing the continuum by breaking it into discrete units, groups, and stages, contrasted with an inspired but sterile holistic view, became dominant in the age of enlightenment and the rational 19th century. The whole emerges from an interaction between parts, and is dismantled iteratively, as with a series of Russian dolls: what is a "whole" on a lower level may be a "part" on a higher level of the hierarchy. However, the distinction between atomism and the continuum becomes blurred on both the microscopic and the macroscopic level. Quantum mechanics ostensibly quantifies reality in some way, but etymology is deceptive here: in fact it blurs discrete particles into fields spreading through space and time; whereas a particle's trajectory is uncertain, the evolution of the wave function is deterministic. The continuum returns on a larger scale when the behavior of numerous individual units becomes irrelevant and science relies on effective continuum theories of thermodynamics, fluid mechanics, elasticity, and electrodynamics. Finally, spacetime itself is a continuum supporting our world just as Atlas used to support the skies – although spacetime may also turn out to be discrete on the unreachably minuscular Planck scale.

The realm of analysis may still be undermined, as the ocean of complexity spreads under the artifice of classification, like the living ocean under a rational space ship in a novel by Stanislaw Lem (1961). In this way, the multiplicity of string theories challenges the neat order of the Standard Model, and the equationless musings of Artificial Intelligence confront established methods of scientific exploration.

1.3 Alphabets and Hieroglyphs

At the base of discretizations and classifications stands an alphabet. The elementary entity is a "letter". Letters combine in "words", words combine in "phrases", phrases combine in "texts". Combinations are governed by "rules", and the result may either have a "meaning" or be senseless. The letters of contemporary fundamental physics are quarks and leptons of the Standard Model, its words are protons, neutrons, and other "fundamental particles" of the mid-20th century, its phrases are atomic nuclei, its texts are quantum-mechanical phenomena. Atoms are the letters of chemistry, its words are molecules, its phrases are polymers, its texts are chemical interactions. The genetic code of DNA has the most concise alphabet of four bases, its words are triplets coding twenty amino acids (Fig. 1.1, left), its phrases are chromosomes, its texts are genetic plans. Cells are atoms of life, its words are tissues, its phrases are organs, its texts are animals and plants. People are social atoms, families are words, tribes are phrases, societies are texts.

An alphabet in its direct meaning (Fig. 1.1, right) was a creation by semitic people of the Eastern Mediterranean who happened to realize that the logic of their language dictated which vowel should be inserted between the consonants of their roots to create an appropriate grammatical form, whereby the syllabic Babylonian script could be simplified to retain bare consonants. The invention appealed to semiliterate practically minded people by its economy, and became almost universal after the Greeks added letters for vowels. But only *almost* universal! The Chinese, cultivating efforts that sustained the promotion of an intellectual elite through hard examinations, held on to their hieroglyphs, which, of course, like letters, are also formed from a finite assembly of strokes, but allow for a much wider variety of combinations.

Nowadays, even the Chinese use a standard alphabetic keyboard to input hieroglyphic texts, supplementing machine suggestions by human memory. But hieroglyphs have another virtue: they have sustained the unity of Chinese script among multiple dialects. If the Romans had adopted Egyptian hieroglyphs rather than the Greek alphabet, Europeans would now be able to read texts in any European language. But it would then be a different civilization. The advantage of Europe was

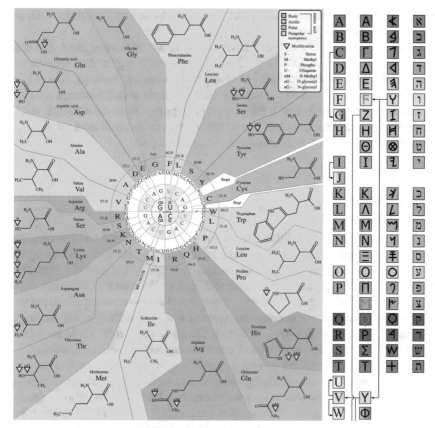

Fig. 1.1 *Left*: The genetic code. From the center outward: bases, codons, and coded aminoacids. *Right*: Latin, Greek, Phoenician, and Hebrew alphabets

in its disunity allowing more latitude for freedom. When a Portuguese king denied funding to Columbus or Magellan, they could turn to a Spanish king, while admiral Zheng He had nowhere to go when a Ming emperor ordered the destruction of his far more powerful fleet. The habit of discerning the constituent parts of speech and transcribing them by elementary entities could naturally evolve into the analytical approach adopted by the Biblical commandments, Greek philosophy, and modern science.

Which kind of a script is preferred by Nature? An alphabetic description of forms works well on the level of physics and chemistry, but blurs as we move higher up the staircase of complexity. Dividing a multitude into discrete partitions might be uncertain. Thus, Linnaean taxonomy becomes murky in assigning the transitional forms of similar species. The sounds of a spoken language depend on continuous shifts in the positions of the lips, tongue, and vocal cords translated into continuous changes of pitch and intensity of sound. They are discretized into a finite number of signs by phonetic transcription, which provides a loose correspondence with the letters of a written language.

Likewise, the hierarchy between DNA and proteins, on the one hand, and cells and tissues, on the other, or between individual human beings and societies is not as straightforward as the one between atoms and molecules. Perhaps, Nature uses hieroglyphic script as well, recognized by different kinds of differentiated cells and driving intracellular dynamics. It may support the unity both of the organism and of terrestrial life as a whole, but it is much harder for us to decipher than the alphabet of elementary particles, atomic structures, and genetic code.

1.4 Mutability

Breaking a continuum into discrete parts brings about stability. Our computers are digital: it is far easier to make a mistake by slightly shifting a number along the continuous scale of an analogous machine than by reversing the state of a binary bit. Alphabetic systems are the most precise and logical; the atoms of Democritus are immutable – but living alphabets are never as precise as their counterparts in physics and chemistry. There are numerous examples of illogical spelling in any language, and first of all in English where, as Bernard Shaw teased, "fish" could be spelled "ghoti" by borrowing odd sound–letter correspondences from different words.

All alphabetic systems are subject to mutations with a frequency that depends both on the strength of an underlying structure and on the level of noise. Moreover, a living system *needs* mutations in order to develop. Mutations can also be deadly, but death is another stage of life. In physics, the measure of noise is *temperature*, and mutability depends on the ratio between thermal energy and the strength of bonds. Applying this back in the language context, we can say that the strength of bonds grows from spoken to written language, and further to established orthography and sacred texts. The temperature here goes up in times of upheaval, invasion, and social change, and down when the language ceases to be spoken.

Bonds between sounds in a spoken language are very weak, although not all sounds are equal. The level of noise is at its strongest when the alphabet, though discernible in the background, is not formalized. Languages evolved and branched, as Steven Pinker (1999) observed, following the same mechanism as the kids game of Broken Telephone. Written language is more stable, to the extent that it would not react to changes in pronunciation, just as English spelling ignored the Great Vowel Shift of the 15th–16th centuries. Stability is further enhanced when orthography is standardized, but new words keep creeping in and the usage of old words shifts. The same Samuel Johnson (1755) who fixed the English spelling used to this day (though not across the ocean) wrote:

> When we see men grow old and die at a certain time one after another, from century to century, we laugh at the elixir that promises to prolong life to a thousand years; and with equal justice may the lexicographer be derided, who being able to produce no example of

a nation that has preserved their words and phrases from mutability, shall imagine that his dictionary can embalm his language, and secure it from corruption and decay.

These words would not stand out if written now, two-and-a-half centuries later, but Shakespeare's language sounds dated, and Chaucer already needs translation.

Language is most stable when frozen in sacred texts, like Latin and Hebrew. Medieval Latin was still apt to deteriorate but Hebrew was rooted in the Bible where every word came directly from God and also had, besides a straightforward reading, a hidden esoteric meaning, coded by numerical values associated with the letters, searched for by kabbalists, up to modern computer-aided studies crowned by an Ig Nobel Prize. The revival of spoken Hebrew by Eliezer Ben-Yehuda in the early 20th century had the effect of an earthquake transforming a landscape, as it opened the language to further changes in Israeli vernacular.

Temperature in its literal meaning has kept falling, on the average, since the Big Bang, just as linguistic temperature was falling, on the average, from hunter–gatherer tribes to modern countries with compulsory school attendance. No structures could survive until nuclear forces became durable enough to keep helium atoms intact against the temperature of one billion degrees kelvin (K), 100 seconds after the Big Bang. Waiting 380 000 years more, the temperature was down to a cool 3000 K, and the plasma soup cleared up, with protons and electrons combining to form neutral hydrogen, whence light broke out. Now, as the cosmic clock shows 13.7 billion years and counting, the average temperature is 2.175 degrees above absolute zero, hardly comfortable for life, but the average is irrelevant, as weather depends on location, up to millions of degrees in star interiors. In the interstellar void, the hell of ice and fire where equilibrium is never attained, the temperature is ill defined, ranging from almost absolute zero to the superstellar energy of cosmic rays.

The temperature in neighborhoods hospitable to life, on planets protected by an atmosphere, is in the range of *chemical* bonds. Their strength is properly measured in units of thermal energy, $E_0 = kT$, where k is the Boltzmann constant and T is the absolute temperature. This measure unequivocally characterizes the capability of a bond to sustain background thermal noise. If E is the energy of a bond, the probability of its break-up is proportional to e^{-E/E_0}. The exponential dependence is very steep. If the energy of a bond is ten times the thermal energy, the probability of its spontaneous break-up at room temperature is less than 0.005%. Typical energies of covalent bonds formed by an electron pair shared by two atoms in a molecule are still more than ten times higher, adding five or six zeroes more after the decimal point. Molecules can be bound by weaker bonds: hydrogen bonds with an energy of about $10kT$ are carried by a hydrogen atom tied to two polar atoms, e.g., oxygen and nitrogen belonging to neighboring molecules; still weaker bonds with energy in single kT units are formed by the interaction of permanent or induced dipoles.

Erwin Schrödinger (1944), in his book pioneering physicists' involvement in the problems of biology, justified at length why the information carrier should be a *molecule*. He followed the work of the geneticists Max Delbrück, later a Nobel Prize winner, and Nikolai Timoféeff-Ressovsky. For the latter, and also the latter's teacher Nikolai Koltsov, this should have been evident, but it was less clear to the physicist accustomed to the idea that macroscopic phenomena should involve a very large number of atoms, since otherwise their ordered structure would be destroyed by random fluctuations. It was already established at the time that genetic traits should be coded by no more than hundreds to thousands of atoms, a minuscule number. In order to keep such a structure intact, strong bonding is needed, and covalent chemical bonds, already understood at the time in the framework of quantum mechanics, were the only way to keep the frequency of mutations low. Schrödinger inferred that a heredity carrier should be an "aperiodic crystal", much more interesting than the periodic crystals commonly studied by physicists. Indeed, only an aperiodic structure can carry information; any sensible text is aperiodic. The double helix structure of DNA, identified by James Watson and Francis Crick (1953) soon afterwards, is indeed both crystalline and aperiodic, with a non-repetitive sequence of base pairs attached to a regular backbone.

Mutation rates generally vary in the range of 1 to 10 per million base pairs across the genome. Within this range of mutability, the genome may evolve on a time scale far exceeding an individual's lifespan. Clearly, not only speciation but replication would become a mess otherwise. Some common antiviral drugs drive viruses into an "error catastrophe" of replication by their mutagenic action (Eigen, 2002). On the other hand, the immutability of gametes on a geological time scale would stop evolution altogether. Rapid mutations in somatic cells are detrimental as well, as they may cause cancer or disrupt a cell's function in another way – but Nature is much less interested in prolonging individual lifespans than in preserving a species in perpetuity, and cancer is as good as any other way to get rid of mortal creatures. The rate of mutation, as it is, is quite right, and it is likely that it has itself evolved to an optimal level in the course of evolution.

1.5 Morphogenesis of Knowledge

Think of a shepherd under starry skies. Think of him at the dawn of history, near the year zero of the Biblical calendar, thousands of years before academia, even thousands of years before learned priests were able to predict solar eclipses. Think of him having the same forceful brain as his descendants in temples and universities in the millennia to come, and having the leisure to stare at the skies – sorry, the pronoun is masculine, his sister has other concerns. What does he see? Constellations of fixed shapes, rising over the horizon, passing slowly along fixed routes, and settling quietly, to rise again the following night, as the Sun and the Moon do in their way. A few wandering stars can be seen among them, but the order of their orbits would also be followed by a discerning eye.

What does he see? He sees heavenly order. He is here because there is order in heaven. Life could not arise, or once arisen could not persist among the chaotic orbits of a double solar system. There is no one looking at a brilliant sunset of a blue sun while the red sun gently warms their limbs. He has no idea that there might be another sun, another sun god, but he knows or rather he feels that the order he sees is created for him and for his tribe, for them to live by this order and to be thankful to its creator. There will be no watches around for thousands of years to come – no watches besides this enormous one in the skies, and no need for them yet, since this one suffices to govern his daily routine and the calendar of his tribe. He is nevertheless sure that there should be One settling the heavenly order, for him and his tribe to live by. He creates his creation myths, thousands of years before the account of God building the heavenly firmament that separates the waters above from the waters below, thousands of years more before it is reinterpreted as a phase transition in the first minutes after the Big Bang. Thousands of years before the all-encompassing waters of Thales and the Pythagorean harmony of numbers, before Aristotelian quintessence resurrecting and fading again in the 19th century aether and the 21st century vacuum, he creates rational myths of the creation of order.

Thousands of years before internet and wide-body jets, and the disorder of globalization – but just a few thousands, not more. There are no negative dates for historic events in the Biblical calendar. His hunter ancestor, with brains as clear and vision as keen, had neither incentive nor leisure to follow heavenly motions. One does not see clear skies in the jungle, nor in the icy storm, nor in the mortal struggle. We think of a shepherd following the motions of the stars by projecting ourselves into the past. Order comes to the heavens when there is at least a semblance of order on Earth. But here, down below, one cannot discern motion as regular as in the skies. True, one can follow the sequence of seasons, one can figure out when it is the proper time to sow and when to drive sheep to higher pastures. The earthly clock is also moved by the Sun. The Sun god dictates social order as well, and tells one to obey the elders. Other things, however, remain unpredictable. Squalls, deluge, drought, still worse, marauding hordes emerging from the desert.

Discerning order and explaining its course was the way followed by philosophers and physicists to come. The aim was understanding, expressing knowledge first in human and then in mathematical language. The latter's power is in its transformational abilities, creating new knowledge by manipulating symbols. The driving force was the pursuit of knowledge *per se*. There were no practical incentives prior to the 20th century; the great advances of mechanics and thermodynamics in the 18th and 19th centuries played a marginal role in the Industrial Revolution. The developing forms of knowledge were still of the same kind that applied to heavenly order. Newton understood logically what the ancient shepherd dreamt about, and it turned out that tangible things on the Earth, like his fabled apple, obey the same rules as the stars.

Both methods and incentives changed when science approached earthly complex systems. Experiments unveiled microscopic structures (with the prefix *micro* mutating now to *nano*) and complex interaction networks. What is the proper form of knowledge in this age? Is it an abstract structure disconnected from observed reality and driven by the pure logic of the mathematical ivory tower? Is it a qualitative model using traditional tools and discerning (given luck) essential features of complex phenomena? Is it a detailed computation elucidating (given luck) a particular detail in a narrow field? Or is it just whatever is funded, cited, and helps academic promotion in a particular discipline? All these forms are pursued with varying degrees of success and satisfaction.

The great mathematician Henri Poincaré said: *The researcher must organize; one does science with facts as one builds a house with stones; but an accumulation of facts is not a science, just as an accumulation of stones is not a house.* The first of the above research modes does not need stones but builds etherial houses in the air. There were architects whose great imaginative projects were never realized, like Étienne-Louis Boullée in the 18th century and Yona Friedman in the 20th, and there is an entire modern school of *conceptual architecture* that is not meant to be built. The second mode is characteristic of 20th century physicists venturing into a general theory of complex systems and biological phenomena. The third mode, merging into the fourth, is becoming prevalent in our day, not by intention but rather by default, as it is hard to find a solvable but unsolved problem of general significance, and experiments are becoming ever more penetrating and precise.

More often than not, stones fail to assemble into a habitable house. The modern shepherd, whatever their (take care of the pronoun – no sexism any more!) actual occupation happens to be, needs earthly science in order to care for their flock, whatever it is. The shepherds of today have little leisure to watch the skies; when free of productive work, they would rather watch a flickering screen, large or small. Which form of knowledge is most useful in practice? Not necessarily generative knowledge, of the kind that brings about new knowledge in the future. And what should be called knowledge, after all? Can an answer produced by an Artificial Intellect be counted as *our* knowledge?

Chapter 2
Aggregation

2.1 The Origins

Inorganic forms evolve according to simpler rules than life, language, or society but create unending variety as well. We turn now to shapes in the non-living world before delving into deeper waters.

Gravity is intrinsically unstable. Right up until the very end of the 20th century it remained uncertain whether our Universe will expand eternally or eventually collapse in a Big Crunch. Its ultimate fate depends on the balance between gravitational attraction and kinetic energy driving the expansion. Neither is known for sure, since they are largely hidden as "dark matter" and "dark energy", but the unexpectedly accelerating expansion of the Universe, or at least of its observable part where we happened to reside, was proven by the efforts of two competing groups who carefully measured the redshifts of supernovas in galaxies lying billions of light-years away, and found them to increase more slowly than linearly with distance.

The global tendency is, however, irrelevant for what is going on at a particular location. One can become rich when everybody around is poor; indeed, if matter is conserved, enrichment is only possible by impoverishing one's neighbors, and this is exactly what the gravitational instability is doing. Local gravitational collapse, a "little crunch", starts when a cloud of gas and dark matter is sufficiently massive and compact, so that free fall toward its center cannot be supported by the gradient of pressure that increases toward the denser inner regions. When exactly it would happen, depends on local energy, both visible and dark, and it is the former, measured by local temperature, which is more relevant, since it should be far more sensitive to local fluctuations. Static equilibrium of a cloud is unstable: it either compacts or disperses, though compaction does not need to go all the way to a final crunch – sucking into a black hole – at least, not straightforwardly, as stable structures may be formed on the way.

There was little growth of perturbations when the Universe was dominated by radiation, but small density fluctuations, about a thousandth of a percentage point, present at the moment when the Universe became transparent, which we still see

© Springer Nature Switzerland AG 2020
L. Pismen, *Morphogenesis Deconstructed*, The Frontiers Collection,
https://doi.org/10.1007/978-3-030-36814-2_2

Fig. 2.1 An image of a spiral galaxy taken by the Hubble Space Telescope

as weak inhomogeneities in the cosmic background radiation, were sufficient to seed the sharp density contrasts that shaped the new world dominated by matter. Whatever we say about the ill effects of extreme wealth and poverty in the social context, the Universe would be boring, dull, and unsuitable for life if it had not developed sharp density and temperature contrasts between stars, planets, and the interstellar void.

Density inhomogeneities develop in a hierarchical way, from galaxy clusters to galaxies, to stellar systems, to stars and planets. The order in which this hierarchy was created is not evident: it depends on the original spectrum of temperature and density inhomogeneities and on the speed with which perturbations with different wavelengths were growing. One possibility is a "top-down" scenario with widespread denser areas giving rise to gas clouds condensing into galaxy clusters, which break up into individual galaxies and further into still denser clouds where stars are born. The opposite "bottom-up" scenario would be a faster growth on short scales, with small denser clumps assembling due to gravitational attraction to larger entities. The young Universe likely became structured in a top-down way, as indicated by a correspondence between the extent of inhomogeneities in the background radiation and the size of galaxy clusters.

In the beginning, even denser regions would still have been "without form", and the first morphing would have been invisible, as matter, falling inward and cooling, left behind a wider halo of dark matter, which would not have cooled due to the lack of interactions sucking up its energy. Free fall would make the collapse complete if motion was strictly centripetal – but velocities of gas molecules and cloudlets are haphazard; just as a satellite does not fall onto Earth when it has a sufficient orbital velocity (depending on its elevation) and planets do not fall onto the Sun, gas clouds start rotating around the central core, and this gives rise to flattened galactic disks. Any uniform distribution of matter also remains unstable on shorter scales. Density waves propagate through the rotating disk, and, since rotation is slower in outlying regions, they bend into spiral arms spreading out of the central bulge (Fig. 2.1).

Spirals are ubiquitous. We will encounter this form in completely different contexts, created by totally different mechanisms, from galaxies down to hurricanes and on to molluscs and plants. Rotation is ubiquitous in a different way, prevailing on the cosmic level, and downscaled only by humans inventing the wheel. Rotation is robust due to the law of momentum conservation, but spirals are fragile: when two galaxies collide, the result is commonly a globular galaxy. Even though the volume within a galactic disk is almost totally void, gravitational interaction between stars disrupts the ordered structure.

2.2 A Star is Born (and Dies)

The intragalactic volume is almost as void as intergalactic space because gravitational instability did not stop at the formation of galaxies, but went on to the birth of the first stars hundreds of million of years after the Big Bang. Gas clouds were never quiet, swept by supersonic turbulence, shaped by shock waves into intermittent filaments and sheets enhancing density contrasts and seeding the cores of future stars.

Although stars lie outside the course of this narrative, they are of the utmost importance for further morphogenesis. They are not just the source of energy for life, but kilns where the chemical elements necessary for life have been forged. Stars have a life of their own, from birth, to maturity, old age, and death, and generations of stars evolve not unlike population cohorts. Before the first stars were born, matter consisted solely of hydrogen and helium atoms. The principal elements of living matter – carbon, nitrogen, and oxygen – are a product of nuclear synthesis within mature stars. Still heavier elements, like most metals, are formed when massive stars explode.

Generations of stars differ because the chemical composition of the Universe evolved as a result of nucleosynthesis in earlier stars. The birth of early stars was the most difficult. Counterintuitively, matter needs to cool before heating up, since internal pressure preventing gravitational collapse grows with temperature, leading to a larger critical mass. The weakly interacting lighter elements of the early Universe cool more slowly, so early stars had to be more massive. The compaction stopped when the temperature had risen enough to ignite nuclear reactions. How could this have happened? We have reproduced nuclear fusion here on the Earth (a potential cause of our own annihilation) by using nuclear fission to ignite it. But there were no heavy nuclei to split in the compressing gas clouds. We are trying to create a tiny artificial sun in a more delicate and manageable way, an endeavor long hoped to come to fruition within the next twenty years (the time span in which Nasreddin Hodja had promised to teach a donkey to talk). Pierre-Gilles de Gennes once said: "We want to put a sun in a box, a nice idea. The problem is how to make the box". So far, all attempts at confining plasma while heating it to attain net energy gain have failed.

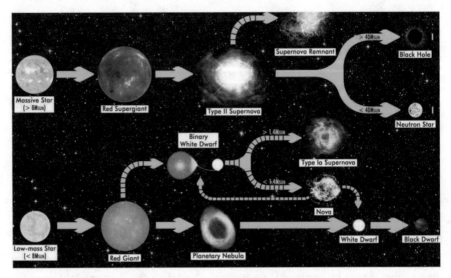

Fig. 2.2 A chart of stellar evolution

For the Universe, making a box was never a problem: the surrounding emptiness is a perfect box. As diffuse matter compresses, the Big Bang scenario goes into reverse. The gas cloud becomes opaque, trapping its radiation. With temperature rising, the hydrogen atoms are ionized, restoring the plasma state that had existed before the Universe had opened up to release what we now see as cosmic background radiation. In the ionized plasma, the magnetic field goes rampant, contributing to its confinement, just as scientists here on the Earth are trying to keep the plasma in check within magnetic traps, in the hope that it will stably heat up till atomic nuclei begin to fuse. In a fledgeling star, this happens as trapped heat reverts it to the state prevailing a few minutes after the Big Bang, when primordial nucleosynthesis was going on. It might go all the way back, to a singularity – a black hole – but time reversal never repeats the past in every detail. Nuclear reactions in star interiors do not repeat those in the early Universe, and they change as stars age and star generations follow each other.

Strictly speaking, Nature never achieves what we are trying to do – to make a safely operating and controllable fusion reactor. In the long run, stars either explode or collapse. The life of massive early stars is "brutish and short", paraphrasing what Hobbes said of early humans. When they turn into red supergiants and explode as supernovae leaving behind a neutron star of nuclear density or a black hole (see the upper part of the diagram in Fig. 2.2), this violent death facilitates the formation of more modest but longer lasting later-generation stars by releasing heavy elements into galactic space. Low-mass stars also age and die (lower part of Fig. 2.2), either with a bang, exploding as novae when they are relatively large or have a close partner in a binary system, or with a whimper, exhausting their nuclear fuel, expanding as red giants and dispersing much of their material to leave behind a white dwarf, and eventually retiring as brown or black dwarfs. Thankfully, the life of a run-of-the-mill star like our Sun is a long run indeed, lasting billions of years. If the mass of the compressing cloud is too small, it may age into a brown dwarf in its cradle, or remain a cold failed star.

Contrary to those beliefs placing paradise in Heaven, outer space is where the most scorching hell is found. While the spiral branches of galaxies are relatively peaceful (occasional explosions excepted), denser star populations in their inner regions collapse into monstrous black holes, up to billions of times the solar mass. The sight of a black hole is anything but black; their environments are the brightest objects in the Universe, illuminated by streams of accreting matter that convert into radiation a far larger fraction of its mass than even stellar nuclear fusion can achieve.

2.3 Sticky Particles

Back on our main track, we turn to the forms created by aggregation. Scaling down from galaxies, we stop on the way to have a look at entities still exceeding our measure – planetary systems. Cold debris, the leftovers of star formation, remained orbiting the central body as it was warming up, collecting in a thin disk (Fig. 2.3, left). The formation of planets from nebular gaseous clouds, not unlike galactic material on a far larger scale, was proposed as early as in the 18th century, first by the Swedish mystic Emanuel Swedenborg, then expanded by the young Immanuel Kant (1755), before he delved into the philosophy of ethics and metaphysics, and in a more detailed form by the great physicist and mathematician Pierre-Simon Laplace (1796), who famously retorted to Napoleon that he had no need to invoke a God hypothesis in his *Système du Monde*.

Universal theories, extending over widely separated scales, from the Solar System, still unexplored, to galaxies, still totally unknown, are characteristic of the 18th century but not the 21st, which takes a more focused look and finds the detailed

Fig. 2.3 *Left*: Image of a protoplanetary disk around a very young star (a youthful hundred thousand years of age). *Right*: Artist's image of a distant hypothetical solar system, similar in age to our own

mechanisms to be far more complicated (see, e.g., Thommes, 2008). There is certainly a clear difference between the formation of gas giants, Jupiter and Saturn, on the one hand, and stony terrestrial planets, on the other. The former condense from gas clouds, mostly hydrogen and helium, in the outer reaches of the protoplanetary disk in a manner similar to star formation, but they are too small on the stellar scale, and fail to ignite nuclear reactions. The material building up inner terrestrial planets is of another sort, comprising dust and small particles, rich in heavier elements. The difficulties in understanding the development of planetary systems are exacerbated by lack of evidence: there is only a single example open for study (and only by observation rather than experiment). It is likely that the majority of stars have planets, some of them, mostly gas giants, already detected, but there is little hope of obtaining elucidating information from faraway stars, unless we may consult alien astrophysicists, if such are ever found.

Gravitation has little effect on dust and small particles but they are apt to collide in a turbulent mass of debris rotating with different speeds. Some of them stick together and are further compacted. Gravity becomes an important factor for planetesimals grown to about a kilometer in size, which will start aggressively accreting smaller chunks. The situation becomes violent when collisions involve larger protoplanets – these are even called "oligarchs", a reference to post-communist *nouveaux riches* competing with similar abandon. Just such a collision between Earth and a Mars-size planet caused our Moon to be born. Then, in a hundred or so million years everything quietened down in our neighborhood, or almost so. Two asteroid belts remain, one placed between the regions of the gaseous and stony planets, where planetesimals failed to aggregate, being perturbed by Jupiter, and another at the outskirts of the Solar System.

Is it just a chance arrangement that we observe in the only planetary system we know in detail? Probably so. The space within a few billion kilometers around the Sun, and very likely of other stars, is full of bits and pieces of all kinds and sizes (Fig. 2.3, right), occasionally bumping into mature planets, sometimes with dire consequences, like the meteorite that wiped out the dinosaurs and another that may wipe us out, unless we are vigilant enough to detect and deflect it. In the long run, as mathematicians have proved, the entire planetary system is unstable, being driven to chaos by small perturbations due to gravitational interactions between rotating bodies, large and small. Thankfully, this "Arnold diffusion" acts on a time scale far exceeding the lifespan of the Sun, but in binary stellar systems, instabilities develop much faster, so it is unlikely that life could evolve there.

While we can only observe the motion of planets, asteroids, and comets and speculate on their past and future, the aggregation of small particles can be studied in the laboratory. It is best done in quiet conditions, in a suspension of colloidal-size particles sticking to each other as the suspension coagulates (Fig. 2.4, left). When coagulation is fast, the structure is ramified, with branches protruding into the surrounding fluid. There is a good reason for this, which is also important on a molecular scale when crystals grow (Sect. 3.6): a branch protrudes in the direction of supply of diffusing particles. With time, the structure becomes more compact, as coagulated particles attract each other, and if aggregation is slow, a branched struc-

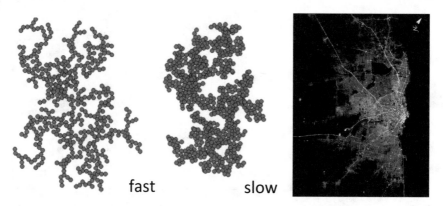

Fig. 2.4 *Left*: Typical shapes of clusters resulting from fast and slow particle aggregation. *Right*: An urban agglomeration (a satellite view of Buenos Aires)

ture thickens before it has enough time to develop. Compare it with urban sprawl (Fig. 2.4, right). Looking from above, we can see branches protruding along incoming roads, a signature of the built-up area expanding in directions of easier commute and/or absorbing satellite towns and villages, while the older central core is compacted.

Is there a connection with planet formation? In the early stages, dust particles were kept together by chemical adhesion, though they aggregated in quite a chaotic environment. Compacting also followed aggregation during later stages when gravity became the major force. This is the reason why smaller heavenly bodies commonly have irregular shapes, while large planets are almost perfect slightly oblate spheroids; the radius of the Earth is about 750 times greater than the height of Mount Everest.

2.4 Power Laws

In all the above examples we see that aggregation leads to extreme inequality. There is no law establishing the average size of a star or a planet or any aggregate or agglomeration. Unlike a normal (Gaussian) distribution where the frequency falls exponentially with growing deviation from the average, the frequency distributions of this kind follow a power law going as $f(x) \sim x^n$ with different exponents $n > 1$. Both kinds of distribution of the frequency of an arbitrary variable x are plotted in Fig. 2.5 (left). It is really hard to show them together, they are so unlike. The power distribution diverges as x approaches zero, so that it should be cut off (or replaced by another distribution) as x goes down to zero. At the other extreme, it has a long tail, and for better clarity it should be plotted in logarithmic coordinates, as has been done in the inset of the figure on the right-hand side. We see in this plot that the power tail attenuates far more slowly than the exponential tail of the normal

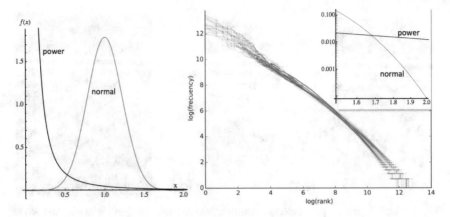

Fig. 2.5 *Left*: The normal and power distributions. *Right*: A log-log plot of the rank versus frequency for the first 10 million words in different languages compiled from 30 Wikipedias. *Inset*: A comparison of the tails of the normal and power distributions on a log-log scale

distribution. Therefore, while there is no chance of meeting a man three meters high, objects or events on the far end of a power distribution are not nearly as improbable. Even though they are rare, they are not vanishingly rare, and are often dominant: strikes of giant meteorites, earthquakes at the far end of Richter scale, world wars, stock market crashes – the *Dragon Kings* of Didier Sornette (2003).

Power laws are ubiquitous because they are characteristic of objects, events, or processes lacking a definite form and strict causality. Aggregation is one such processes, whether it relates to the formation of planets, or to accumulation of capital, or to the growth of cities (although in the latter cases, subject to social interactions, deviations from a power law indifferent to all detail may be substantial). Power laws are also often formulated by taking as a variable x the *rank*, i.e., the position in a certain list instead of a physical measure. This is done in the law of distribution of the frequency of words (Fig. 2.5, right), rather undeservedly called Zipf's law. George Kingsley Zipf attributed this tendency to laziness: people prefer to reach understanding in the easiest way. Unlike some other denizens of power law tails, rare words are not threatening. Good writers and skilled speakers carefully choose uncommon words to express their thoughts and intensions most clearly.

Power laws are never precise. We can see in the right-hand panel of Fig. 2.5, and indeed in any other power law plot, that straight lines in the log-log coordinates change their incline and become fuzzy at both ends. There is a good reason for this. These plots are just empirical, compiled by plain counting, and this is where errors are felt more strongly and specific features of a particular set of data interfere. At the high end, where only a few of the most frequent words are present, only a few words are counted, and their frequencies are influenced by the grammars of particular languages. Far down at the low end, the distribution is influenced by the sporadic appearance of rare words. It is in the middle part that all differences are blurred, where both the number of words and their frequencies are large, and statistics is most precise, and it is from there that the prevailing exponent of the power law is extracted.

In other applications, where the variable x is a physical property rather than a rank, the log-log plot is drawn in another way, with larger sizes or stronger events in the tail and frequent ones at the high end, but the cause of deviations is the same: there are very few large planets, strong earthquakes, and world wars, while on the other hand small cosmic debris, hardly felt tremors, or minor skirmishes may escape attention. The reader can say at this point that objects obeying a power law have no place in a book on morphogenesis. They are here for the sake of contrast: "universal" laws indifferent to content, compiled statistically, and offering no sure predictions is all we are left with in a world without form. Another reason is that we should pay respect to Dragon Kings defying forecasts. Disasters may drive progress. We would very likely not be here now if dinosaurs had not been put down by a Dragon King.

2.5 Phase Transitions

Liquid droplets aggregate still more readily than solid particles. Why have I not yet mentioned liquids? Both liquids and gases are *fluids*, and both are without form, differing only by density, and smoothly merging one into the other at a critical point. Although we clearly distinguish between air and ocean here on the Earth, the distinction would not be so clear-cut elsewhere. There is a lot of water vapor in the atmosphere. What would happen if the oceans vaporized, as they probably did on Venus? Is the interior of stars gaseous or liquid or neither? It is a plasma, of course, since it is ionized, and this property trumps distinctions in density.

The separation between a liquid and its vapor is just one example of a *phase transition*, encountered in many other settings, from magnetic spins to separated neighborhoods. The equilibrium state of a physical system is determined by the minimum of the free energy, $F = E - TS$. Molecules of the same kind attract each other, and the energy level E is lowered when they come close together. But the entropy S is at its highest level when they occupy all the available space, which they would always do if they do not interact, as in an ideal gas. As the temperature T grows, entropy prevails, and the phases will not separate above some critical temperature. This kind of a transition between two disordered phases differs qualitatively from solidification, which is always abrupt, as it establishes a certain order, even if imperfect.

We have to go down to a milder climate, closer to what we are able to endure, to see liquid and vapor as distinct coexisting phases. They are separated by a sharp interface carrying extra energy, and therefore tending to minimize its area. Molecules within the bulk of a liquid interact on all sides with molecules of the same kind, while molecules near the surface have agreeable neighbors only on one side (Fig. 2.6, left). If this is the boundary with the vapor phase, there are few molecules of the same kind there. If this is the boundary with another fluid phase, interactions with molecules on the other side are less favorable – this is the origin of *surface tension*.

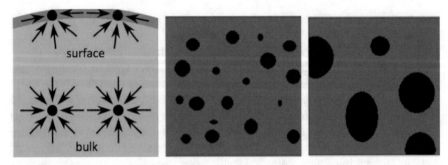

Fig. 2.6 Surface tension. *Left*: Forces acting on a molecule in the bulk of a fluid and on its surface. *Center to right*: Matthew's law for droplets

For the same reason, droplets suspended in air or in another fluid acquire a spherical shape to reduce their energy; coalescing droplets swiftly round up, far more readily than dust aggregates. On the other hand, the upward pressure of concave troughs is able to support a certain weight; water striders take advantage of this to walk over the surface of a pond. The optimal shape can only be distorted by gravity or flow in the surrounding medium, and a lot of effort is needed to spray or spatter a bulk liquid. Moreover, the more strongly curved a convex interface, the fewer friendly neighbors molecules have on the liquid side, and the more easily they evaporate or dissolve. This is the Kelvin effect (Thomson, 1871). It causes the *ripening*, or coarsening, phenomenon: smaller droplets wither away and larger ones grow at their expense (Fig. 2.6, from the center to the right) – recall Matthew's law (Sect. 1.1).

Surface tension decreases as the critical point is approached, phase separation then becomes imperfect, and density fluctuations grow. This causes the fluid to become opaque sufficiently close to criticality, when the size of fluctuations becomes comparable to the light wavelength. Although phase separation driven by minimizing free energy is easily understood, it is far more difficult to deduce its dynamics and near-critical behavior directly from molecular interactions. A number of toy models have been suggested for this purpose. The most popular one, motivated by magnetization, was invented by Ernst Ising (1925). There are two opposite orientations of tiny magnets (or particles with opposite spins) placed on the nodes of a two-dimensional grid. This is a square grid in the left-hand panel of Fig. 2.7, but it could be a different one. The magnets interact with their closest neighbors, and switch their direction if it lowers their energy. Ising himself did not think that a phase transition would be possible in this system, but a phase transition does indeed take place and, as the interaction strength grows, all the little magnets orient in the same way, so that the grid is "magnetized". The toy model became important when Lars Onsager (1944) solved it precisely, which made it possible to understand in detail the behavior near the critical point.

Models of this kind found their way into applications far removed from physics. Thomas Schelling (1969) suggested a similar model describing segregation of ur-

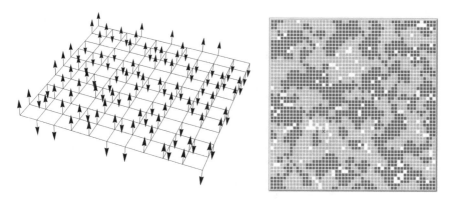

Fig. 2.7 *Left*: The Ising model on a square grid. *Right*: Schelling's segregation model

ban neighborhoods (Fig. 2.7, right). The model is also set on a square grid, and oppositely oriented spins are replaced by agents of two kinds, which are supposed to become uncomfortable when the number of neighbors of another kind exceeds a certain number; some cells are left empty. The model was evidently aimed at racial segregation, and as these agents, unlike spins, cannot change their color from pink to blue or the other way around, they ease their discomfort by moving to an empty cell. Quite naturally, this leads to a phase transition to a segregated state, with blue and pink "races" occupying distinct clusters. What was considered special in this model, and won Schelling a quasi-Nobel prize in economics (Onsager got the real thing – in physics), was a demonstration that segregation can be totally spontaneous, without any collusion or command.

The idea of more general models of this kind, based on agents moving on a two-dimensional grid according to set rules, goes back to Stanislaw Ulam and John von Neumann in the early postwar years, but they wouldn't engage in this seriously. Such "cellular automata", easy to design and implement even on the computers of yesteryear, have been used to design a plethora of games which bear little relation to reality but are capable of generating all kinds of dynamic patterns never reaching equilibrium. The first and most famous of them was John Conway's *Game of Life* (Gardner, 1970). Steven Wolfram (2002) attributed reality to such games, proclaiming them "a new kind of science", to the bewilderment of mainstream physicists expecting illuminations of another kind from the former young genius.

Chapter 3
Broken Symmetry

3.1 Crystals

Symmetry breaking, though exterminating unlucky creatures in its cruel ways, often operates in a more refined and benevolent manner by creating *structures*. The laws of the Universe are built in a way allowing matter to *self-organize*, enhancing its complexity at the price of increasing entropy elsewhere. Atoms gather to form molecules, or they assemble in crystals; molecules combine in polymeric chains and colloidal particles, polymers evolve to self-replicating circuits, and colloidal particles evolve to cells, and these keep self-organizing to form animals, plants, ecological landscapes, and societies. On all scales, symmetry breaking turns uniform spaces into *patterns*.

Crystals are periodic patterns on the atomic scale. Atoms or molecules assemble in a suitable order to optimize their interactions. The order is counteracted by *entropy*, defined by the formula chiseled on the grave of Ludwig Boltzmann, its creator: $S = k \ln W$, where the Boltzmann constant k multiplies the logarithm of W, the number of possible realizations of a certain state. The perfect order is unique, the logarithm of unity is zero, and so is its entropy. On the contrary, ways to disturb the order are plentiful. There are many ways to remove an atom from its due position in a perfectly ordered crystal, and this number is huge in a crystal visible to the naked eye. There are still more ways to place a stray atom in the interstices of the crystal grid, disturbing its neighbors. Perfect crystals require strong interatomic bonds and special conditions for their creation; therefore diamonds are not only (almost) forever, but also very expensive. Thermodynamics teaches us that equilibrium structures strive to minimize their free energy $F = E - TS$, where E is the internal energy depending on the strength of the interaction bonds and T is the temperature, which, multiplying entropy, favors disorder. As a consequence of this simple formula, crystals melt when the temperature rises. This was known long before Boltzmann, and the very term *crystal* stems from *kryos* ($\kappa\rho\upsilon o\varsigma$), the Greek word for frost.

Perfect order may come in different varieties. The order is rational, and was already precisely understood in the 19th century. The crystal structure is determined

---

The original version of this chapter was revised: Figure 3.5 has been replaced. The correction to this chapter is available at https://doi.org/10.1007/978-3-030-36814-2_9

© Springer Nature Switzerland AG 2020

L. Pismen, *Morphogenesis Deconstructed*, The Frontiers Collection, https://doi.org/10.1007/978-3-030-36814-2_3

by the structure of a unit cell, infinitely repeated by translation in three dimensions. There are 32 possible crystal classes. Johann Hessel (1831) derived them, without the benefit of mathematical group theory, by studying actual crystal forms. This classification, published in an obscure Physics Dictionary, went unnoticed until reprinted posthumously in 1897. Auguste Bravais (1850) proved the existence of 14 Bravais lattices in three dimensions, based on 14 types of unit cells that can be repeated by translating in all three dimensions to build up all crystalline structures. Additional crystal classes can be obtained by rotating unit cells.

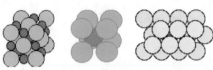

Fig. 3.1 Ionic lattices with the coordination numbers 6 (*left*) and 8 (*center*) and a metal lattice with the coordination number 12 (*right*)

It is far more difficult to predict the shape of a unit cell and hence the type of lattice formed by a particular chemical species: this depends on the arrangement of atoms and a preferred orientation of bonds that would optimize their interactions and reduce the energy of the crystal. Even apparently similar crystals with a simple composition, like pure metals composed of identical atoms held together by delocalized electrons or ionic crystals containing ions of opposite sign may settle into different unit cells. In both cases, the energy minimum can be attained in different compounds with different coordination numbers, i.e., different numbers of adjacent metal atoms or ions of the opposite sign, which interact attractively. For ions of different sizes, like Na^+ and Cl^- in common table salt (NaCl), the optimal three-dimensional ionic arrangement excluding contacts between mutually repulsive ions of the same sign has the coordination number six – along the three mutually perpendicular directions in a cubic lattice (Fig. 3.1, left). Ions of about the same size can be packed more efficiently with the coordination number eight, with each ion sitting in a body-centered cubic cell surrounded by the corresponding counter-ions (Fig. 3.1, center). For identical metal atoms, the maximum coordination number with twelve neighbors is possible (Fig. 3.1, right).

Structures held by covalent bonds are more varied, since such bonds have a certain direction. Carbon atoms commonly have four covalent bonds directed to the vertices of a tetrahedron. This is the basis of the diamond structure (Fig. 3.2, left).

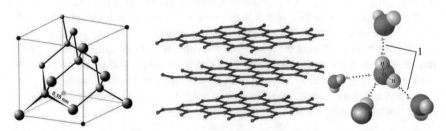

Fig. 3.2 The structure of diamond (*left*) and graphite (*center*). *Right*: Hydrogen bonds (shown by *dotted lines*) in ice

Within graphite layers, chemical bonds are similar to those in a benzene ring, already somewhat weaker, forming a flat hexagonal lattice (Fig. 3.2, center), but layers are bound only by van der Waals bonds, the weakest of all. Water molecules in an ice crystal are held together by weaker hydrogen bonds due to a hydrogen atom shared by two oxygen atoms. The configuration of the ice crystal (Fig. 3.2, right) depends on the shape of the water molecule, and this is the reason why ice is, rather uniquely, lighter than water, to our (and marine life's) benefit; denser crystalline structures of ice do exist at lower temperatures and higher pressures, thankfully under conditions far removed from our (and marine life's) everyday experience.

Clearly, the strength of bonds determines the melting point of a crystal, as we already noted at the beginning of this section, and ice held by hydrogen bonds melts at a far lower temperature than most salts and metals. Diamond and graphite melt or sublimate at a temperature above 4000 K. Far higher melting temperatures are possible beyond common earthly conditions. Van Horn (1968) predicted the crystallization of ions at temperatures of millions of degrees kelvin at enormous densities and pressures in the interior of white dwarfs, and this has recently been confirmed by observing the decrease in their cooling rate due to the release of the crystallization heat. Crystals at still higher densities and temperatures are sustained by nuclear forces in the core or in neutron stars (Glendenning, 2001).

Crystalline structures can be obtained by superposition of standing waves of the form $a_k \cos k \cdot x$, where x is the vector defining the location and k is a *wave vector* of a suitable magnitude and direction. The wave vectors characterizing a particular structure are revealed by the Fourier transform of the spatial density and/or the charge distribution. This is the basis of standard tools for studying crystal structure: diffraction of X-rays or electrons with a wavelength of the same order of magnitude as the interatomic distances. The energy of a crystal can be presented as a sum of *resonant* combinations of waves, such that their wave vectors sum up to zero. The simplest resonant combination includes triplets of waves with the same wavelength. In two dimensions, there is only one such combination with wave vectors forming a regular (equilateral) triangle, which corresponds to the hexagonal pattern filling a plane. In three dimensions, such resonant combinations may form one of

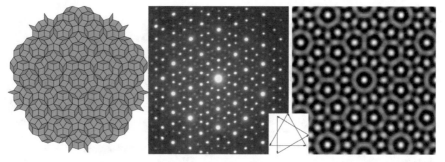

Fig. 3.3 *Left*: Penrose tiling. *Center*: Electron diffraction pattern of an icosahedral quasicrystal. *Right*: Quasicrystalline structure obtained as a superposition of six waves directed as shown in the *inset*

the Platonic bodies with triangular faces: a tetrahedron, octahedron, or icosahedron. The former two correspond to the face-centered and body-centered cubic crystal lattices, but the latter is something special, as we shall see presently.

Of course, other resonances are possible, including more than three waves and waves of different wavelengths leading to other crystal structures, but the above-mentioned lattices are the densest and most common, in both two and three dimensions. By classical theory, periodic crystals may possess an n-fold rotation axis with n equal to 2, 3, 4, or 6. When Dan Shechtman (1984) found the forbidden 5-fold symmetry in a diffraction pattern, he hesitated to publish. His hesitation was superfluous, and the discovery eventually led to his 2011 Nobel Prize in Chemistry. Before Shechtman's experiments, quasiperiodic structures had been predicted by Harald Bohr (1925)[1] as projections of regular crystals in higher dimensions, and in the 1970s Roger Penrose invented a quasiperiodic tiling of the plane bearing his name. This pattern, shown in the left-hand panel of Fig. 3.3, never repeats itself but has a simple structure built up of just two types of rhombic tiles. A quasiperiodic structure in the plane obtained by superposition of six waves forming two resonant triangles shifted by 30° looks similar to the diffraction pattern of a quasicrystalline material (Fig. 3.3). A three-dimensional quasicrystal is generated by the icosahedral resonant structure.

The crystalline structure shows up in the shape of slowly growing crystals, as we see in the shapes of snowflakes (Fig. 3.4, center) retaining the hexagonal symmetry of the crystalline lattice of ice (Fig. 3.4, left). A quasicrystal may grow to a dodecahedral form (Fig. 3.4, right), which is impossible when the classical symmetries are obeyed. Not only simple molecules, but also proteins and nucleic acids may crystallize. The standard tool for studying crystal structures – X-ray diffraction – was instrumental in understanding the conformations of biological macromolecules, and in particular, in establishing the double helix structure of DNA (Sect. 4.2) .

Most solids, except carefully grown crystals, are polycrystalline. This may affect their mechanical properties in different ways. On the one hand, decreasing the number of grain boundaries makes gliding along atomic layers easier and thereby makes a metal softer. On the other hand, polycrystalline structures are more brittle, as they are apt to fracture along grain boundaries. Admixtures have a similar effect;

Fig. 3.4 *Left*: The hexagonal crystalline structure of ice. *Center*: Snowflakes. *Right*: A dodecahedral quasicrystal

---

[1] The mathematician brother of Niels Bohr.

thus, bronze is stronger than pure copper, as was understood when new kinds of tools and weapons initiated the Bronze Age. Even single crystals are weakened by *dislocations* (more on them in Sect. 3.4). Ways to attain better order may be counterintuitive. Reheating a crystal may remove imperfections, as the system is teased out of a metastable equilibrium. In this way, steel is tempered and particle arrays are stirred up to better compact.

3.2 Crystals on the Mesoscale

The structure of crystals is determined by interactions between atoms, which can be both attractive and repulsive, as seen most clearly in ionic crystals. Atoms and molecules (or even people) preferring neighbors of their own kind tend to separate (Sect. 2.5), but aggregation of separated phases on larger and larger scales is prevented if they are tied together. This happens in block copolymers which include blocks of mutually repulsive units, like AAAABBBBBAAAABBBB. Failing to separate completely, such polymers form patterned structures, not so neat as crystals but approaching an ordered structure on a mesoscopic scale that depends on the size of the blocks (Fig. 3.5).

The characteristic inner scale of structures of this kind is very different from that of atomic or molecular crystals. It is called the *mesoscale* – in-between the atomic and macroscopic scales. It could literally be called microscopic, as it lies just below the micron in most applications, within the range of an optical microscope. However, the prefix *micro* has always been reserved for the tiniest things – a scale that kept decreasing over time as science penetrated deeper into the makeup of matter. Other labels are colloidal or *nanoscale*, referring to *colloids*, suspensions of particles with sizes from just a few to a few hundred nanometers.

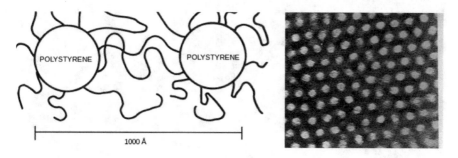

Fig. 3.5 *Left*: A scheme of separation of block copolymer units. *Right*: The resulting pattern

Better ordered structures of this kind are called *colloidal crystals*. They may have the same structure, most commonly the densest, that is, face- or body-centered cubic as in an atomic crystal, but consist of particles of a colloidal size. A natural mineral of this kind is *opal*, a close-packed array of silica spheres. What makes opal a gemstone is that the internal scale of a colloidal crystal is of the same order of magnitude as the wavelength of visible light. Just as X-rays are diffracted by atomic crystals, thereby revealing their structure, so is light when it shines on opal, giving it either an *opalescent* appearance or brilliant colors.

Changes in lattice spacing affect the transparency and reflectivity with respect to light at a certain wavelength, which earns them yet another appellation – *photonic* crystals. Animals took advantage of this before us. Chameleons change their color, either for camouflage or for social signaling, by modifying the lattice spacing of a nanocrystalline layer that covers their skin, as in the left-hand panel of Fig. 3.6 (Milinkovitch, 2015). Some butterflies owe their brilliant colors to similar layering on their wings. They are not smart enough to manipulate it, but another insect, the Hercules beetle, can do this (Hinton and Jarman, 1972). The transparent cover of its wings is underpinned by a spongy layer containing an array of pillars normal to the surface (Fig. 3.6, right). The beetle may fill this layer either with air to make it yellowish or with water to turn it black.

In technology, the Hercules beetle's trick is used in optical sensors. Here the aim is not just to change color but to investigate the properties of a fluid filling the porous space of a mesoscopic crystal. The structure, called a *reverse opal*, is prepared by first assembling monodisperse silica or polymer spheres in a close-packed lattice, and then filling the interstices with another material and removing those spheres to create an ordered porous medium. The void can be filled by a fluid

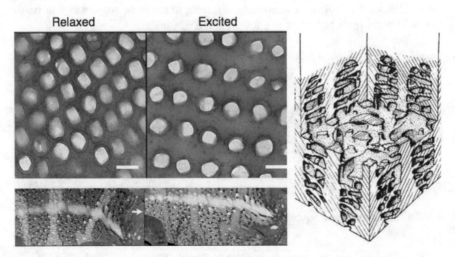

Fig. 3.6 *Left*: Change in the lattice spacing (*top*) and the resulting color change of a chameleon in a relaxed and excited state (*bottom*). *Right*: Side view of an array of pillars in Hercules beetle wings, cut in the middle to show the top view

that we seek to test. A soft filling material will change its volume depending on the fluid's composition, ionic strength, or some admixture of biomolecules, which, in turn, will change the color of the transmitted or reflected light (Stein et al, 2013). In a similar way, the material in the interstices of a colloidal crystal may respond to all kinds of factors, besides the chemistry of the imbibed fluid, including electric field, temperature, or mechanical stress. It is not necessary to remove the particles to obtain chameleon-like optical effects due to swelling and shrinking, as in the upper left panel of Fig. 3.6. This was proposed as a way to construct a full-colour digital display (Arsenault et al, 2007). Our TV and computer screens are not built in this way, at least not yet, since displays based on the change in polarization of liquid crystals (LCD) or electroluminescent diode (LED) are currently more practical.

Octopuses and their cephalopod relatives, squid and cuttlefish, create optical effects in still more sophisticated ways. Rather than carrying a permanent photo-crystalline layer, they create it on the go. Their skin has arrays of *papillae*, small rounded protuberances containing pigmented cells – *chromatophores* – controlled by muscles, which can dynamically modify their size and shape (Allen et al, 2013). In addition, their skin has light-sensitive elements, and parts of the brain are distributed in their tentacles; this distributed sensing and actuation contributes to extremely fast action. Modern technology imitates it in the design of flexible camouflage sheets that can autonomously sense and adapt to the coloration of their surroundings (Yu et al, 2014).

3.3 Liquid Crystals

When a solid melts, its crystalline order is lost but some liquids retain vestiges of order. These are *liquid crystals*. The appellation sounds like an oxymoron: crystal implies order, and liquid, fluidity, but matter may acquire intermediate forms possessing elements of order while still being fluent. The most subtle vestige of order is a preferred alignment of molecules that depends on their chemical structure and shape. This is a *nematic* order, it may arise when molecules are elongated and stiff,

Fig. 3.7 Schemes of liquid-crystalline phases. *Left to right*: nematic, cholesteric, smectic A, smectic C, and smectic C*

usually due to some rigid elements, like double bonds and benzene rings. The nematic alignment is characterized by a *director*, which is like a vector without an arrow, as shown by the spindles in Fig. 3.7. Circular symmetry around the director may be lost; then the nematic becomes *biaxial*. Nematic order may emerge not only on a molecular scale, but also on a mesoscopic scale. Cells can be polarized in anisotropic tissues (Sect. 6.4) and oblong bacteria also tend to align nematically (Sect. 7.3).

The sketches in Fig. 3.7 show other ordered liquid phases. In the *cholesteric* phase, the prevailing orientation rotates around some axis. In *smectics*, molecules are layered; within each layer, they are oriented normally in the A phase, at a certain angle in the C phase, at at an angle rotated between one layer and the next in the C^* phase; within each layer, their positions are disordered. As in all phase transitions, order diminishes as temperature grows, so that, for example, a solid may melt into a smectic, which melts into nematic, and then to an isotropic liquid.

Liquid crystals possess a kind of elasticity which, unlike in solids, does not hold back deformations but only works to orient them uniformly. In nematics, the change of orientation can be of three kinds: splay, bend, and twist, as sketched in the three left-hand panels of Fig. 3.8. Since there is no translational order, nematics can flow like normal fluids, only their viscosity is anisotropic. Smectics are already sensitive to the distortion of the shape of their layers and are only mobile within the layers.

If an isotropic liquid is "frozen" into the nematic state, the orientation will be different at different locations, and there will be a lot of *defects* – points where the orientation becomes indefinite. Defects are not just arbitrary flaws: they obey precise laws of *topology* (Kleman, 1983). The topological *charge* of a defect is measured by the rotation of the director along a surrounding contour. Stable defects with the lowest energy have the lowest possible charge. Since the director is invariant under rotation by $180°$, defects may have charge one-half, either positive or negative. Paired defects of this kind in the plane are shown in Fig. 3.8. This picture can also be viewed as a cross-section of a half-charged line defect in three dimensions. If such a line forms a closed contour, it can collapse into a point defect, called, for obvious reasons, a *hedgehog* (see the rightmost panel of Fig. 3.8). Defects of opposite signs attract each other. If the pair of defects shown in Fig. 3.8 merges and annihilates, the alignment will become uniform, at least locally, but it is impossible to eliminate all defects if the nematic orientation is required to remain at a certain angle to the boundary or on a closed surface, like a sphere.

From a practical point of view, what is special in liquid crystals is their optical properties. Due to their anisotropy, liquid crystals polarize light. Light, as a transverse wave, has two polarization directions, and a liquid crystal transmits the one aligned with its director; it also rotates the polarization of light when oriented at some angle to a light ray. The orientation of the director is easily controlled by treating the confining surfaces and using electric or magnetic fields. In the pixels of standard displays, the light beam passes through a polarizer that polarizes it parallel to the orientation of the liquid crystal, set by treating its surface. The orientation of both the polarizer and the liquid crystal at the other end are turned through $90°$, and light, rotated by the twisted liquid crystal, passes through. When a voltage is turned on, the liquid crystal orients along the beam direction, the polarization of the light is not rotated any more, and the beam is blocked by the adverse polarizer at the exit. Of course, the nematic alignment within such pixels has to be uniform.

Sometimes liquids solidify without becoming more ordered. Glass and carbon-based elastomers are disordered, *amorphous* solids. When glass melts, there is no abrupt transition, it just gradually becomes more malleable, which is convenient for molding its forms. Essentially, solid glass is a liquid with a very high viscosity that decreases as the temperature rises. Elastomers are entangled polymer chains interacting only through weak intermolecular forces. They are *viscoelastic*, which means that their response to deformation is intermediate between that of solids and liquids. Rubber is hardened by cross-linking these chains. Molecules of liquid crystals can be polymerized, either connecting "head to tail" (main-chain) or being attached to a polymer backbone (side-chain), as in the two upper left-hand panels of Fig. 3.9. They then turn into soft solids, retaining their partial order – *liquid crystal elastomers*.

The creation of such materials, envisaged as artificial muscles, was an idea of a theorist, the winner of the 1991 Nobel Prize in Physics, Pierre-Gilles de Gennes

Fig. 3.8 *Left to right*: Splay, bend, and twist distortion of nematic alignment; a pair of half-charged defects of positive (*top*) and negative (*bottom*) sign; a hedgehog defect

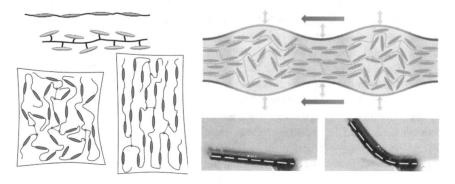

Fig. 3.9 *Top left*: Schematic structure of main-chain and side-chain nematic elastomers. *Bottom left*: Working principle of an artificial muscle: reshaping due to a phase transition from the isotropic to the nematic state. *Top right*: Peristaltic locomotion of a worm by travelling waves of radial expansion and longitudinal contraction. *Bottom right*: Bending of a nematic elastomer plate due to one-sided actuation

(1975). When this material is "actuated", prompting a phase transition from the isotropic to the nematic state, it elongates along the director and accordingly shrinks in the normal directions to preserve its volume, as shown in the two lower left-hand panels of Fig. 3.9. It goes the opposite way when the transition is reversed. Elongation is stronger for main-chain elastomers. In this way, a liquid crystal elastomer can really work as a muscle, lifting a considerable weight, but much more can be accomplished through reversible phase transitions under the action of light, temperature, or chemical agents.

In the example shown in the upper right-hand panel of Fig. 3.9 (Palagi et al, 2016), a wave enforcing a transition from the disordered to the ordered state and back makes this nemato-elastic "worm" crawl in the direction shown by the arrow. If the phase transition is enforced on one side only, as in a layered material or when only one side of a flat strip is heated or illuminated, the strip bends, as shown in the lower right-hand panel of the same figure (Camacho-Lopez et al, 2004). Strings, sheets, or shells of liquid crystal elastomers can be made to walk and swim, and even used to construct soft robots – more on this in Sect. 8.2.

3.4 Non-Equilibrium Patterns

Symmetry breaking takes place on a macroscopic scale in systems sustained far from equilibrium by external fluxes. Broken symmetry is most evident in fluid mechanics. Water and air are never still, and neither are fluid-like granular media forming sand ripples and dunes. The Earth itself, fluid on a geological time scale, breaks its spherical symmetry by its ever changing rugged relief. While we enjoy varied landscapes and the play of waves, we have to distinguish between symmetry broken by outside interference, when waves are caused either by wind or local perturbations, like throwing a pebble, and *spontaneous* symmetry breaking – self-organization, which can be inferred in a pattern of cloudlets we often see when looking up to the skies or down from an airplane window. A hurricane is a powerful solitary structure spontaneously emerging from humidity above warm ocean waters and winds and currents driven by the rotating Earth.

The basic scheme of self-organization of this kind, repeated not only in hydrodynamics but in various non-equilibrium processes, is shown in Fig. 3.10 (left panel). We start with a featureless plane. The external input is directed perpendicularly, without disturbing the symmetry – but the result is a *pattern*. Since the vertical direction is reserved for the external flux sustaining the system in a non-equilibrium state, the emerging patterns are two-dimensional. The most common pattern is a hexagonal grid, generated by the resonant interaction of three waves forming a regular (equilateral) triangle. It can be distorted, with differently oriented patches separated by strings of defects, as in the central panel of Fig. 3.10. Striped patterns can be seen as well, usually distorted to a labyrinthine tangle, sometimes coexisting with hexagons, as in the right-hand panel of Fig. 3.10.

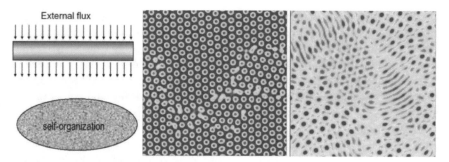

Fig. 3.10 *Left*: Scheme of pattern formation under the action of a uniform external input. *Center and right*: Distorted patterns in simulations of convection with a deformable interface

Fluid mechanics was the arena for the first controlled experiments that demonstrated non-equilibrium patterns originating in spontaneous symmetry breaking. The first experiment of this kind was carried out by Michael Faraday (1831) who observed standing surface waves in a vertically oscillating fluid or sand layer. The next experiment, the one which became most famous, was carried out by Henri Bénard (1900), who observed hexagonal convection cells in a thin layer of whale oil heated from below. The mechanism was explained by Lord Rayleigh (1916): the light warm fluid raises upward, cools there, and descends. Rayleigh carried out linear stability analysis and computed the critical temperature difference across the layer, above which convection starts. The theory turned out, however, to be wrong in respect to Bénard's actual experiments, where similar convection patterns were caused by surface tension gradients driving the fluid along its free surface from locations warmed up by ascending currents to cooler patches. The convection patterns are similar, and the term Bénard–Marangoni convection distinguishes it from the gravity-driven phenomenon, which is more common, and is the only possible one when the fluid is confined between two solid plates.

In chemical applications, awareness of instabilities causing spontaneous pattern formation had to wait another half-a-century till the famous work of Alan Turing (1952), philosophically charged and winged by the fame of the Turing machine and the Enigma Code. The paper bears the ambitious title *The chemical basis of morphogenesis*, but ends on a humble note: *It must be admitted that the biological examples which it has been possible to give in the present paper are very limited. This can be ascribed quite simply to the fact that biological phenomena are usually very complicated.* It is not about biology at all, but about chemical patterns, and the rational message to be extracted from the 36 long pages is that pattern formation requires, in the simplest setting, the combination of a slowly diffusing activator with a rapidly diffusing inhibitor. This principle, which can be established in a few lines by the linear stability analysis of a two-component reaction–diffusion system, is prominent in many model pattern-forming systems.

The way a Turing pattern is formed is schematized in Fig. 3.11. If the level of the activator is raised locally (upper panel) it also raises the level of the inhibitor (middle panel). Since the latter is more diffusive, it spreads out to an area where the activator level is low, depresses the activator there, and, as the latter's level goes down, is depressed itself (lower panel). An ecological analogy would have herbivorous animals attracted by local abundance of grass, then trampling it underfoot. The undulations propagate out, leaving in their wake a pattern with a wavelength that would typically be about the geometric mean of the diffusional ranges of each species.

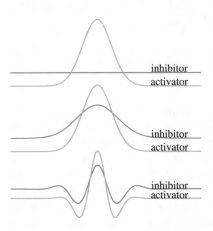

Fig. 3.11 Scheme of patterning involving a slowly diffusing activator with a rapidly diffusing inhibitor

Both activator and inhibitor can be either chemical or biological species, or, in a more abstract form, other physical agents. For example, in Bénard convection, driven either by gravity or surface tension gradients, temperature plays the role of an activator, and fluid viscosity, the role of an inhibitor. The scale of the pattern is determined by the relevant spreading ranges. In desert vegetation patterns (Meron, 2018), plants serve as an activator and the *lack* of ground water sucked in by plants as an inhibitor. Pictures of patterns of different origin can hardly be distinguished, but the final result may depend on the geometry of the region, on some special interactions, or just on inhomogeneities and initial conditions. With time, patterns may evolve into more regular shapes, e.g., as dislocations collide and annihilate, but this process, similar to coarsening, is very slow, and often enough, defects cannot be eliminated for topological reasons.

The tug-of-war between structuring forces and entropy continues on all scales. Special care is needed to preserve regular patterns. It is very difficult to manufacture a perfect grid of the metamaterial described in the preceding section, able to create optical effects or even bend light rays in bizarre ways, or slow them down to standstill. It is next to impossible to create a perfect information-bearing *aperiodic* crystal: a book or a genetic code with no misprints, or social order with no infractions. Special agents may be employed to sustain order, from correctors to antibodies to police.

Crystalline order can be disrupted in different ways defying precise classification, like the formation of vacant sites (holes), atoms or ions drifting into the interstices of the crystalline grid, or admixtures of impurities. If all this is avoided, perfect order is still challenged by *topological defects*. Jan Burgers (1940) described crystal dislocations in terms of the failure of a contour, followed along the edges of a fixed number of unit cells, to close. The distance and direction between the start and end points is defined by the Burgers vector; it is parallel to the contour for an *edge*

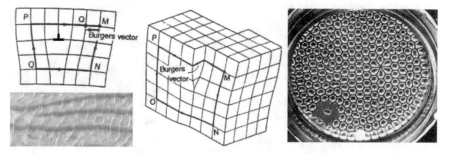

Fig. 3.12 An edge dislocation (*left*) and a screw dislocation (*center*). *Right*: Photo of Bénard's original experiment with a penta-hepta defect accentuated by colors

dislocation (Fig. 3.12, left), and perpendicular for a *screw* dislocation (Fig. 3.12, center).

The edge dislocation is retained in two-dimensional striped patterns. Just count the number of ridges crossed when one goes around the frame on the lower left-hand panel of Fig. 3.12, with the plus sign when moving up and the minus sign when moving down. Again, we have a unit Burgers vector. A striped pattern is, strictly speaking, not a crystal: it is patterned only in one direction and, defects excluded, homogeneous in another one. Hexagonal planar patterns contain *penta-hepta* defects: adjacent cells with five and seven neighbors. One such defect is accentuated in Bénard's convection pattern in the right-hand panel of Fig. 3.12. Chains of such defects are seen along the boundary between patches with different orientations in the central panel of Fig. 3.10.

3.5 Oscillations and Waves

Patterns of another kind, which are possible only in systems sustained in a non-equilibrium state, are *dynamic*. In fluids, all kinds of dynamic wave patterns are typical, while stationary patterns of the kind mentioned in the preceding section (where, of course, the fluid still moves within convection cells) need special arrangements. Oscillatory patterns, like stationary ones, need the combination of an activator with an inhibitor. The simplest model of this kind was the prey–predator model devised by Alfred Lotka (1910) and later independently by Vito Volterra. The prey is an activator: it multiplies by itself, and makes growth of the predator's population possible. The predator is an inhibitor, depressing the prey's population and unable to multiply on its own. The original Lotka–Volterra model is defective, as it has a continuum of oscillatory solutions with different amplitudes depending on an analogue of "energy" conserved on each orbit. This non-generic feature is brittle, and disappears when the equations are modified to take into account any realistic effect, like saturation of the prey's growth at higher densities.

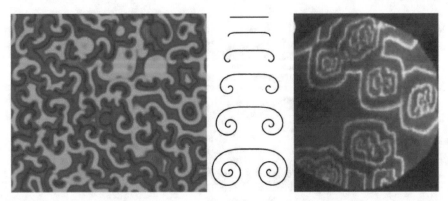

Fig. 3.13 *Left*: Snapshot of a disorganized spiral wave pattern in the BZ reaction. *Center*: Scheme of the formation of a spiral wave. *Right*: Anisotropic patterns on a catalytic surface

The situation with chemical patterns is actually similar – standard equations of chemical kinetics, like standard prey–predator equations, contain just products of concentrations of different species. Nevertheless, chemical oscillations were considered taboo because of thermodynamical misconceptions well into the 1960s. Turing (1952) did not consider this possibility either. There was some reason for this prejudice: chemists were used to reactions in a closed vessel, rather than in continuously operating reactors with reactants supplied and reaction products removed, which can operate under non-equilibrium conditions. Electrochemical oscillations and waves had been known since the turn of the 20th century (Ostwald, 1900), but there an electric current clearly sufficed to prevent relaxation to equilibrium. Chemical oscillations were discovered in 1951 by Boris Belousov, but the prejudice was so strong that he was barred from publishing it, and only in 1959 managed to squeeze a note into an obscure collection of medical abstracts. Belousov observed oscillations in a closed vessel, and they did indeed come to a halt when at least one basic reactant was exhausted – but continued for a sufficiently long time to be clearly seen.

A young graduate student Anatol Zhabotinsky was encouraged by his supervisor to find out how Belousov's recipe worked. Zhabotinsky (1964) observed not only oscillations but fascinating target (circular) and spiral wave patterns in a Petri dish. The Belousov–Zhabotinsky (BZ) reaction became high fashion in the 1980s, but its chemical mechanism is still unknown in all its details, and other oscillatory reactions have since been discovered. Oscillations are common (though usually not welcome) in exothermic chemical reactions, where temperature plays the role of an activator and exhaustion of a reactant is an inhibitor.

Oscillations naturally transmute into propagating waves when the extent of the reacting medium is large compared to a characteristic diffusional range of the reactants – practically always, unless the liquid is stirred. Spirals are a generic wave pattern: a target wave initiated at some point turns into a spiral under any perturbation; putting a finger into a Petri dish of Zhabotinsky's low-tech experiment is enough. The central panel of Fig. 3.13 shows how a spiral is likely to form. The edge of a front segment propagating upwards in the picture lags behind and, as the segment spreads out, spirals are formed at its sides. Patterns and waves on a much finer scale (visible under an electron microscope rather than by naked eye), and with an added anisotropy distorting round wave fronts to a squared form (Fig. 3.13, right), were observed on catalytic surfaces and earned Gerhard Ertl the 2007 Nobel Prize in Chemistry. Their mechanism is more complex, as it involves restructuring the catalytic surface.

A typical pattern is a maze of spirals, as in the left-hand panel of Fig. 3.13. A wave is characterized by its propagation direction, which is a vector rotating by 360° around the point wherefrom a circular or spiral wave originates. This point is therefore a defect of unit charge – compare with the half-charge defects in the alignment of a director lacking an arrow (Sect. 3.3). The charge is everywhere positive: as you go round a contour, the vector rotates in the same direction. Defects of the same charge repel each other, which prevents the ordering of a disorganized pattern.

The formation of self-organized patterns, either stationary or oscillatory, is such a basic and widespread phenomenon that its essential features can be imitated even by a very simple model. This is the FitzHugh–Nagumo (FN) system (FitzHugh, 1961), first derived as a simplified version of the nerve conduction model by Hodgkin and Huxley (1952), which is itself a far-reaching simplification. What is essential is to have two reaction–diffusion equations: first, a nonlinear activator equation, and second, an inhibitor equation, which can be linear. A cubic nonlinearity suffices to have two levels of the activator that would be stable at some level of the inhibitor concentration.

The way this system works is illustrated in Fig. 3.14. The net formation rate of the activator vanishes on the S-shaped curve in the plot spanned by the activator and inhibitor concentrations, while the inhibitor concentration grows to the right and decreases to the left of the inclined straight line. The only stationary state is the intersection of the straight and S-shaped lines in the center, and it is unstable. If the activator level lies on the left branch of the S-shaped curve, the inhibitor decays until its level reaches the bottom of the S-curve. Beyond this point, a stationary activator level cannot be sustained, and it rises fast to reach the right-hand branch. Now the inhibitor concentration goes up until it reaches the upper bend of the S-curve, the activator level drops back onto the left branch, and so it goes. This temporal oscillation translates into a wave pattern when its phase changes from place to place. The two inhibitor levels indicated by the arrows in the picture are then translated into the front and back of the activator wave. The same system can generate stationary patterns of the kind discussed in the preceding section. The boundary between domains with the two concentration levels of the short-range activator is a stationary

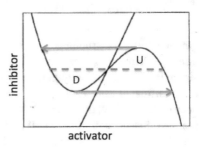

Fig. 3.14 Scheme of the activator and inhibitor dynamics in the FN model

front, placed on such a level of the long-range inhibitor (marked by the dashed line in Fig. 3.14) that the areas U and D are equal.

The FN system and models of this kind have been applied to a variety of pattern-forming processes, from the BZ reaction to nerve conduction to ecological interactions to coloration of animal fur, sometimes reflecting the actual mechanism in a simplified way, sometimes producing apparently similar patterns in spite of having no connection with reality. When the real world is recalcitrant, we build ourselves toy models based on equations simple enough for us to solve. Sometimes a toy model provides illuminating insights into behavior in the real world. More often, it remains what its name implies, a plaything for mathematically inclined physicists (Dyson, 2004).

The possibility of oscillations on a microscale, labelled "time crystals", was proposed by Frank Wilczek (2012). This was followed by a string of papers proving what should be quite clear: that no time-periodic state is possible without an external drive. Quantum effects notwithstanding, oscillations cannot take place at equilibrium. "Time crystals" were realized in a laboratory (Lukin et al, 2017), but only when sustained by a periodic drive. The period of oscillations is an integer multiple of the driving period, so that oscillations of such "time crystals" are not autonomous. A "space-time crystal", that would be seen when oscillations propagate as waves, is a still more fancy term sounding as if it is connected to relativity theory, although it does not touch upon it in any way.

3.6 Branching Patterns

Another common structure, besides stationary or dynamic patterns, seen in various non-equilibrium systems, from phase transitions to living forms, is the *dendritic* structure. This is typically generated by growth. If a crystal grows by accretion of atoms diffusing from a solution, its flat surface is unstable. If there is slight protuberance on the surface, this will be better exposed to the solution, and will grow further. On the other hand, a dimple will be less accessible and will be left further behind. The farther a protuberance grows and the closer it comes to the source of material, its sides will lose stability and it will start to branch out in its turn (Langer, 1980). The result is a dendritic structure like the one in the left-hand panel of Fig. 3.15. Some bacterial colonies are structured in the same way, as they grow in response to the supply of a nutrient (Ben-Jacob et al, 1994).

This instability never saturates, the interphase boundary remains unstable, and a dendrite keeps branching and growing. Of course, this goes for an ideal dendrite in an infinite medium. In reality, any instability saturates at some level that is not accounted for in the basic model. The growing crystal will slow down when its tip becomes too sharp, close to the molecular scale or, in the case of a bacterial colony, to the size of an individual microbe.

Branching patterns also can be caused by hydrodynamic or elastic instabilities. Viscous fingering develops through the Saffman–Taylor (1958) instability in

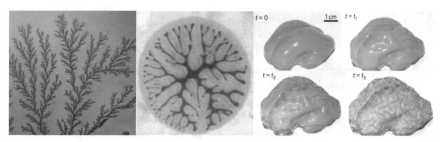

Fig. 3.15 *Left*: Dendritic copper crystals. *Center*: Dendritic flow pattern in a Hele-Shaw cell. *Right*: Time sequence for the development of a strongly buckled surface in the simulated growth of cerebral cortex during gestation

a Hele-Shaw cell, a shallow space between two parallel plates where a liquid can be pumped. If a less viscous fluid displaces a more viscous one, a protuberance on the interface encounters less resistance and grows further, creating a fingering pattern like the one in the central panel of Fig. 3.15. Fingers are slowed down and kept from further branching by surface tension – but before this limit is reached, a highly branched pattern can develop.

Elasticity more strongly constrains this kind of instability, but the surface of a soft body buckles as it grows. Tallinen et al (2016) observed and simulated a soft gel coated with a thin layer of elastomer gel that swelled when immersed in a solvent. The volume grew more near the surface, which gradually became more and more convoluted. The aim was to simulate the growth of a human brain during gestation. The inner gel mimicked the white matter, the outer layer represented the cortex, and the initial shape was molded by the image of a smooth fetal brain. The time sequence of forms in the right-hand panels of Fig. 3.15 is very similar to the progressive formation of the cerebral cortex with its cusped *sulci* and smooth *gyri*. Branched and convoluted structures are ubiquitous in Nature. We will meet them again in growing plants and blood vessels (Sect. 7.1), as well as in nerve networks (Sect. 7.2).

What is special in branching patterns is their extremely large surface area (or boundary length in two dimensions). In this way, it approximates a *fractal* object. If you measure the length of a convoluted line, like the boundary of either the growing or the fingering pattern in Fig. 3.15, using a ruler and placing it in such a way that both its ends lie on the line, the total length increases as you decrease the length of the ruler, with a power equal to the line's fractal dimension, which will come out to be between one and two. The same will happen when measuring the area of a strongly buckled surface, leading to a fractal dimension between two and three. Again, this increase is limited by physical factors affecting the fine structure of the pattern: a fractal object is a mathematical concept that can only be approximated in the real world, something rarely mentioned in popular renditions (e.g., Mandelbrot, 1982).

3.7 Self-Organized Universe

Think again of a shepherd at the dawn of history (Sect. 1.5). Under the skies with their orderly revolutions, his earthly surroundings are full of mysteries, bristling with entangled forms and surprising occasions. He does not try to analyze them, he takes them for granted. How could this infinitely complex web have arisen, with its landscapes, plants, animals, and people? Certainly, not by itself. It might have been created by a mighty God, the same one that set the heavenly clock in motion – or had it always existed? The biblical account of creation (Sect. 1.1) is straightforward; it lacks sophisticated philosophical insights, and this is why it can be seen as an unartful paraphrase of the modern story of creation.

The monotheistic story becomes questionable when one delves more deeply into its details. Even if God is infinitely mighty, micromanagement is not an efficient strategy for the head of a large enterprise. The shamanic world view, with a host of local spirits responsible for winds, waters, forests, and landscapes in their infinite and changeable variety, could appear more natural, and not by chance it precluded any unified outlook. This tradition remained more deeply rooted in the East, culminating in *karma* or *Tao* as vaguely defined concepts of the spiritual roots of cause and effect determining the natural order of the eternal Universe. Lacking a supreme manager, the world had to be self-organized,

Thinkers tangential to modern science are apt to link the ideas of self-organization in complex systems far from equilibrium developed in the mid-20th century, also dubbed in a more fancy way *autopoiesis* (self-creation), with the Eastern wisdom (Capra, 1975; Lent, 2017). The connection is, however, merely superficial. The God of the monotheistic tradition does not need to be involved in minute details, does not need to take care of the way leaves are growing and blood flows in veins. It is sufficient to set up laws and let them follow their course, which even the Creator would not be able to predict. Complex systems are sensitive to minute details of the environment and initial conditions and defy precise forecast, but this does not mean that they do not meticulously obey physical laws discovered, verified in minute details, and widely applied by scientists, great and humble. The complex forms of the Universe, or at least of the universe we are living in, both beautiful and ugly, both dangerous and benevolent, came into being through self-organized evolution; and broken symmetry, going far beyond the basic examples in this chapter, was a major mechanism for creating the endless variety we witness in the world around us. Monotheistic origins fade into the image of God beheld by Spinoza and Einstein, merging with the Laws of Physics.

The holistic attitude leads nowhere. It doesn't allow us to penetrate into the inner workings of Nature. Self-organization is driven not by neo-Confucian *li* imposing rational order on the raw material force of *qi*, but by specific mechanisms determined by physical laws, even though different causes may generate similar forms. Activators and inhibitors have nothing to do with the complementary opposite forces of *yin* and *yang*. The analytic, versus holistic, approach, reflecting the contrast between rugged Judean and Greek landscapes and the expanse of China, and mirrored in the contrast between first feudal and then democratic discord, and imperial unity, led eventually to the contrast between advances triggered by the Renaissance and the Industrial Revolution and the misery of China in the 19th and 20th centuries. The decline of the East did not have intrinsic roots in the mystic Eastern soul: this was promptly proven by first Japan, and then India and China embracing Western science and technology and often surpassing their originators. China and South Korea have won all the more recent Physics Olympiads, while a photo of the US team was full of East-Asian faces, with only a single white boy.

Neither has holistic philosophy led to a harmonious social order. Taoist ideas of harmony with Nature never brought China "self-regulating homeostasis", which *may hold valuable lessons for our current global society*, as Jeremy Lent (2017) writes, expressing weighty and genuine concern for the fate of the accelerating hamster wheel of world economics. I doubt he would like to live there and then. Chinese "homeostasis" periodically collapsed in violent turmoil as dynasties changed. Their medieval

Fig. 3.16 Chinese philosophical symbols

stories often end with the protagonist put to death along with fifty of his relatives after being defeated or incurring the emperor's wrath. European stories of the time are also full of violence – but families are spared.

Where we would live if the holistic attitude prevailed? In a world with no more than a billion people, with a life expectancy at birth of thirty years, with no chemical plants polluting the atmosphere, no cars, no planes, no carbon dioxide emissions, no nuclear weapons, no internet, and no hackers (unless in the literal sense), and nothing known about the world within and above us, nothing to write about in a book like this one. A systemic outlook encompassing the web of interactions in complex systems is necessary – but any action affecting their course should rely on the knowledge of specific mechanisms. Crystallization of order is inevitably tied up with the emergence of defects, and further evolution at a more refined level should come in to heal their adverse impact, as when tempering steel.

Chapter 4
Life Evolves

4.1 Alive or Not?

We can follow the ways life evolved and try to understand, slowly entering into the details, the way it operates, but the toughest questions are *how* life emerged and *what* is life. The latter question is more philosophical than scientific: *a philosopher's job is to find out things about the world by thinking rather than observing*, declared Bertrand Russell (1972) – and with neither observation nor mathematical insight one cannot get very far. Ludwig Wittgenstein (1922) was wise at this point, observing that *the solution of the problem of life is evident at the disappearance of this problem.*

In a pioneering study of this hard problem, Alexander Oparin (1924) listed metabolism, self-reproductivity, and mutability as defining properties of life. What is sufficient – any of these or all of them? A virus replicates but needs the facilities of an infected organism or cell to do so, and it lacks self-sustained metabolism. Is it alive? Some answer "yes" and some "no". A computer program (or a computer virus) replicates – is it alive? Hardly. A robot can manufacture other robots. Even a book replicates in some way but, not unlike a virus, it needs a printer. All of the above are mutable, even a book. A fellow philosophical question is: what is consciousness? As it goes with all philosophical questions, they can be and have been discussed for centuries. It was easier once when everything could at the end of the day be attributed to God. There are grey border areas between the black and the white, but our intuition tells us what is certainly alive or aware of itself and what is certainly not.

Traditional notions of the origin of life were rather contradictory. On the one hand, everything was stated clearly: *And God said, Let the earth bring forth grass*, etc., and two days latter *God created great whales, and every living creature that moveth, which the waters brought forth abundantly, after their kind, and every winged fowl after his kind*, and *God said, Let the earth bring forth the living creature after his kind, cattle, and creeping thing, and beast of the earth after his kind*, and finally, before retiring for a good rest on the Sabbath, *God created man in his*

© Springer Nature Switzerland AG 2020
L. Pismen, *Morphogenesis Deconstructed*, The Frontiers Collection,
https://doi.org/10.1007/978-3-030-36814-2_4

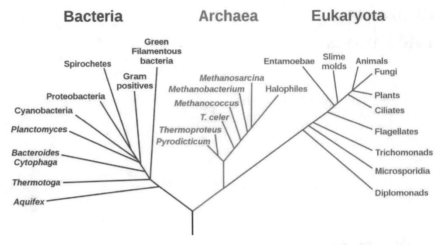

Fig. 4.1 A phylogenetic tree

own image. There is a rough correspondence here with our notion of the order of emergence of different classes of animals as a result of evolution, though whales are lumped with fish rather than mammals, and "creeping thing", whatever it is, should have preceded birds. But look at a modern phylogenetic tree (Fig. 4.1)! Animals and plants are just two small entries in the upper right corner. It would be hard, of course, to pay attention to bacteria before the microscope was invented, and armies of biologists had to be employed to classify the rest of life's "kingdoms" or "empires", as they are labeled in some taxonomic systems, and their subdivisions. In any case, it would be too long a story to include in the first chapter of a popular publication. But why did only sizable animals deserve attention? No insects were taken to Noah's Ark, or at least none are mentioned. Likely they were stowaways. The lower classes earn more attention and respect today. James Lovelock (1979) compares large plants and animals to elegant salesmen and glamorous models, much less important than microbes, the tough workers who keep things moving.

Apparently, underclass creatures did not deserve the honor to be created by God, and beliefs in their spontaneous generation from non-living matter, like flies from putrefying earth and even mice from mud, persisted as late as the mid-19th century. This belief had to be abandoned in the end when Richard Remak (1855) proved that any cell can only be begot by division of pre-existing cells, but apparently what was still missing was the authority of a famous Parisian professor. The final nail in the coffin of spontaneous generation was an elementary experiment by Louis Pasteur (1864) proving that no organisms grew in a properly sterilized and sealed flask.

The Panspermia hypotheses, brought forward in their most coherent form by Svante Arrhenius (1908), and discussed by some great 19th century scientists before him, sought to avoid the question of the origin of life by relegating it, if not to God, to the skies. According to this hypothesis, microbial life or at least its essential pre-biotic organic constituents were distributed throughout the Milky Way, or possibly also among galaxies, by space dust, meteoroids, and even alien space ships, as imagined by none other than the famous discoverer of the structure of genetic code (Crick and Orgel, 1973). This, of course, only pushes the hard question back, redirecting it to the original source. In an extreme interpretation, it would presuppose the unity of life, with the same chemical constituents all over the Universe, notwithstanding dissimilar environments on countless planets that may support life, perhaps based on totally different chemistry, as speculated by biologists, astronomers, and science fiction writers. There is, nevertheless, a persistent interest in a search for traces of life in meteorites, and it is quite certain that microorganisms are capable of surviving in a cosmic environment for a very long time.

The mainstream approach remains, however, earthbound. Charles Darwin imagined a *warm little pond, with all sorts of ammonia and phosphoric salts, lights, heat, electricity, etc. present.* Alexander Oparin (1924), then in his twenties in the still innovative Soviet Union, suggested that gradual evolution of organic chemicals in the Earth's "primordial soup", similar to Darwin's, in an atmosphere rich in methane, ammonia, hydrogen, and water vapor, may have led inadvertently to the emergence of life. He was seconded by John Haldane and John Bernal, both, perhaps not accidentally, communist sympathizers. There was quite some excitement when Stanley Miller (1953) demonstrated spontaneous synthesis of amino acids within a sealed glass flask filled with warm water and a gaseous mixture imitating what was believed at the time to be the early Earth atmosphere, fired by sparks imitating lightning bolts. This supported Oparin's opinion that there was no fundamental barrier between organic chemistry and life.

The question of the actual mechanism of the emergence of life remains nevertheless open to this day, and the "warm little pond" scenario is no longer a leading candidate. Even if the amino acids in Miller's experiment polymerized by chance, forming a clumsy semblance of a protein molecule, this is still a very long way from sustainable life. A complex molecule emerging by a fleeting whim of chemistry will disintegrate in response to another whim, unless it is capable of reproducing itself. Life is a self-organizing process; its elements must join into a structure capable of building and enhancing itself by joining into new elements.

We should expect it to be harder to uncover the origin of life than to understand the way the Universe came into being: we see the distant past of the Universe by observing its far reaches in spacetime much more clearly than we see the past of our Earth, hidden in geology and in the genetics of extant organisms. Moreover, even in the case of a brilliant success, the picture we would see would be merely parochial. Life, at least primitive life, certainly exists on a great many planets, and the particular circumstances of a transition from their various chemistries to their various life forms will not necessarily conform to whatever happened here.

4.2 Memory and Function

The grey period between molecules and morphs may have lasted on this planet for hundreds of millions of years. The scholastic question: which was first – the chicken or the egg – has mutated in modern deliberations on the origin of life into the question: which was first – amino acids, assembling into proteins, or nucleotides, assembling into RNA and DNA (RiboNucleic or DeoxyriboNucleic Acids). The former are working catalysts, the latter are memory elements – but their relations are mutual: proteins catalyze the assemblage of amino acids coded by DNA to form other proteins. John von Neumann (1951), the great mathematical physicist and mastermind of modern computer architecture, addressing the origin of life from his standpoint, compared proteins and genes to hardware and software. In computer technology, the question of which came first does not arise, since both hardware and software are created and evolved by an intelligent designer. A computer lacking software can "metabolize" – consume electricity – but cannot do anything else, and, of course, software without hardware is useless, and would never have been written. Hardware and software were developed in parallel, enhancing both computational capabilities and the convenience of programming hard tasks. In later days, when personal computers became a commodity, this symbiosis became perverse. While latter-day programmers were economical, trying to extract more output from clumsy hardware, modern developers bloat operating systems and applications to make out-of-date computers obsolete and induce us to buy new ones.

The 1967 Nobel prize winner in chemistry Manfred Eigen reinterpreted von Neumann's simile in more abstract notions of *function* and *information*. In his words (Eigen, 1971), *"function" cannot occur in an organized manner unless "information" is present and this "information" only acquires its meaning via the "function" for which it is coding*. Eigen envisaged evolution in the Darwinian sense of "survival of the fittest" taking place through competitive growth even as early as the prebiotic stage, by selecting the most efficient cooperative circuits of autocatalytic reactions that might have become precursors of living matter. Eigen's definitions are more precise than von Neumann's: the carriers of both function and information are *molecules*, life's hardware.

Replicating structures at the boundary between chemistry and life had to be as simple as possible. RNA is both a short-term memory device and a catalyst. In the organisms dominating now – *eukaryota* (Fig. 4.1), DNA base sequences are copied on short RNA strands, which serve as messengers sent from the cell nucleus to the *cytoplasm* where the coded proteins are assembled. However, RNA molecules are capable of replicating themselves, so they can function both as genotype and phenotype – what Eigen (1971) calls "self-instruction". This supported the idea that the earliest life-like forms were based on RNA (Rich, 1962). Tiny plant pathogens – *viroids* – may be a relic of the archaic "RNA world" (Diener, 1989). Viroids are just RNA strands closed to form a loop. They do not even have a protein coat as viruses do, but they are capable of replicating, though, like viruses, they need the right enzyme for this – a protein supplied by the host. Here, too, cooperation between RNA (or its more primitive precursors of which no evidence remains) and proteins

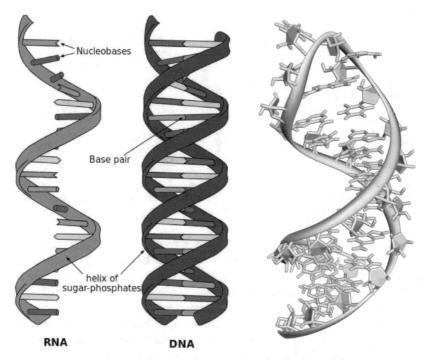

Fig. 4.2 *Left*: A single-stranded RNA compared to a double-stranded DNA helix. *Right*: A single RNA strand folding back and pairing with itself to form a double helix. The nucleobases are shown in *green* and the ribose-phosphate backbone in *blue*

is necessary, and this suggests that they might have co-evolved at the dawn of life. Replication of synthetic RNA has also be observed in the laboratory by Philipp Holliger's group (Pinheiro et al, 2012).

We have already noted (Sect. 1.4) that, while mutability of genetic material is indispensable for evolution, too rapid mutations are detrimental, and that a digital code is less prone to mistakes than a continuous (analogous) representation. The discrete character of genetic information was established through a quiet low-tech study of variation in peas by Gregor Mendel, published in 1866 without anybody noticing it for a third of a century. Mendel essentially discretized heredity. He proved that inherited traits do not mix: white-flowered and red-flowered plants, when cross-fertilized, do not produce pink-flowered progeny. What is inherited, is a dominant trait (in Mendel's experiments, the red color of certain flowers), but the recessive trait (white color) can reappear with probability 1/4 in the offspring of two its carriers (Fig. 4.3, left). Recessive genes often carry harmful mutations; this is why custom and religions prevented incest in most cultures for millennia before Mendel.

Starting in 1908, Thomas Hunt Morgan carried out extensive cross-breeding experiments on the fruit fly *Drosophila melanogaster*, which remains up to this day the most widely studied "model animal". He honed Mendelian laws by detecting

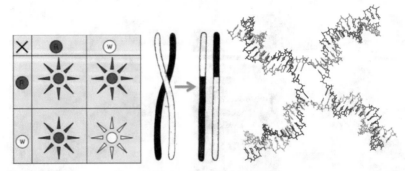

Fig. 4.3 *Left*: A scheme of Mendelian inheritance. *Center*: Illustration of crossing over by Morgan (1916). *Right*: Molecular structure of the Holliday junction that leads to crossing over

the phenomenon of *crossing over* (see Fig. 4.3, center), the exchange of gene sequences between two chromosomes (Morgan, 1916). The swapping of maternal and paternal DNA enhances genetic variation and makes any offspring unique, otherwise the genes residing on a parental chromosome would be inherited all together. The structure enabling crossing over, a knot formed by two partially separated strands of two double spirals, called the *Holliday junction*, is shown in the right-hand panel of Fig. 4.3. The frequency of separation between particular genes due to crossing over could be used to indicate the distance between them along the chromosome; this was a useful tool before modern fluorescence-based sequencing methods were developed.

Another complication, leading to a still higher genetic variation and influencing evolution, is the *horizontal gene transfer* common among single-cell organisms. On these occasions, genes are transferred from a donor species to a possibly unrelated receiver species. This plays havoc with the phylogenetic tree connecting its separated branches to a network of a richer structure (Fig. 4.4), and complicates attempts to establish the origin of the various genetic lines.

It is easy now to forget that the genetics of the first half of the 20th century was advancing in the dark, with a wrong idea about the nature of the carrier of inheritable traits. It was supposed that inheritance resided in a certain kind of particles, which Darwin called *gemmules*, and it was established early in the 20th century that this carrier in turn resided in the chromosomes – but it was assumed that genetic information was contained in proteins. This was quite natural: proteins are ubiquitous in cells, and versatile. The very term, coined in 1838 by the great chemist Jacob Berzelius after the Greek word *proteios* (primary), tells of their primary importance. The true culprit was only identified when Oswald Avery and coworkers induced bacterial transformation using pure DNA (Avery et al, 1944). This quite naturally attracted attention to nucleotides, culminating in the discovery of the double helix structure of DNA by James Watson and Francis Crick (1953).

In retrospect, it should be clear that proteins are unfit to serve as memory elements. They can, in principle, form autocatalytic replication cycles, but mistakes in their composition are unforgivable. Each protein has its own conformations, and modifying its structure even in a minor way is likely to disrupt its assigned function. On the other hand, if a base in DNA is replaced, the structure of the double helix remains intact, and mistakes in replications turn into the driving force of evolution.

The coding genetic alphabet turned out to be minimal and neatly organized: just four bases in a linear sequence (Fig. 4.2) coding in triple combinations all proteins in the living chemical factory. It would be twenty amino acids in a plethora of configurations if proteins were carriers of genes. Eigen (1971) explains this simplicity by a pragmatic argument: nucleation of any code is easiest when the number of digit classes is as small as possible. Yet human alphabets did not follow this principle: the number of graphemes was decreasing rather than growing at the stage of early evolution, and even in counting, the evolution went from the Babylonian sexagesimal figures to our common decimal, to binary code. But who knows? Perhaps the code of the earliest life contained more "letters", abandoned when more efficient memory mechanisms were found.

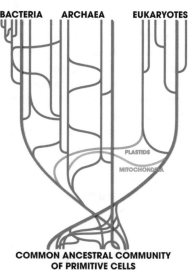

Fig. 4.4 The phylogenetic tree modified by horizontal gene transfer

Single-stranded RNA molecules are unstable, compared to DNA double helices tied together by bridging bases (Fig. 4.2, left), but their stability is enhanced if they fold back and pair with themselves as in Fig. 4.2 (right). This comes out more easily in the closed loop of a viroid RNA. Stability can be further enhanced when two RNA strands couple to form a structure approaching that of DNA. How could it happen? A second RNA genome – identical and apparently redundant – would ensure replication if one strand were damaged. It might enter a primitive cell when it exchanged material or merged with another cell, not necessarily carrying an identical RNA strand, and this mechanism has been suggested as the origin of sex in the very early stages of the evolution of life (Bernstein et al, 1984). It could also be feasible for two separate strands to link up in a double spiral, and in this way RNA world would gradually transmute into our DNA-based biosphere.

Primitive forms could afford a higher mutation rate, and even benefitted from faster evolution, but stability had to be enhanced with the emergence of more complex life forms, which had to possess larger genomes and to persist longer, as both individuals and species. As Eigen (1971) formulates it, *for optimal selection, the required precision of information transfer has to be adjusted to the amount of information to be transferred.* Higher organisms have evolved genome repair mechanisms, which make use of the complementarity of DNA strands: if a segment of one of them is damaged, it can be restored to fit its intact counterpart.

4.3 Alien Earth and Gaia

The planet where life was emerging was very much unlike the planet we live on. It was an alien Earth where hardly any organism we know would survive, as there were only traces of oxygen at best, but plenty of poisonous gases in the atmosphere. This is the reason why the search for the origins of life drives interest toward *extremophiles*, microbial life thriving under unusual conditions: in hydrothermal vents at the Mid Atlantic Ridge, near sulphurous volcanic cauldrons, or buried deep under the rocks, or in the ocean trenches, or in the Antarctic's icy underground lake. Various strains of bacteria may be resistant to radiation, get along without oxygen, or thrive in strongly acidic, alkaline, or salty environments.

Another focus of interest is life on exoplanets; several thousands of those have been discovered at the time of writing (early 2019), and this number is bound to grow to five figures by the time this book is printed. We are unlikely ever to be able to consult alien colleagues, even if they are found somewhere in our close galactic vicinity, just a few scores of light years away, still less collect bacterial samples; but detecting the composition of planetary atmospheres is feasible, and this can give a clue. Erwin Schrödinger (1944) wrote: *Living matter evades the decay to equilibrium.* If the atmosphere contains oxygen and gases that can react with oxygen, for example, methane, this is already a telling sign that one or both of them are of a biogenic origin, and it could be life, even if primitive, that maintains the atmosphere in a non-equilibrium state. Closer to home, primitive life might be possible on Mars, in the higher layers of the atmosphere of Venus, and in the interior of Jupiter's and Saturn's moons.

Palaeontology provides far more information on the early life here on Earth than exoplanets will ever do. Fossilized microorganisms more than four *eons* (billions of years) old have been found in hydrothermal vent precipitates in Quebec, and evidence of early life on land was discovered in 3.5 eons-old mineral deposits in Western Australia. This indicates the very early appearance of life forms, soon after the oceans were formed. Moreover, life may have appeared even earlier than known records indicate, and may have been extinguished and emerged anew. At the time, although the Sun was weaker, the young Earth might have been very hot due to frequent impacts of large meteorites and intense radioactivity, and it has therefore been suggested that the first microorganisms may have been hyperthermophilic, thriving at temperatures above 80°C.

Life needs an energy supply – where could energy come from? The ultimate energy source for life as it exists today is sunlight supporting the photosynthesis of organic matter by plants, algae, and cyanobacteria, with oxygen released as a waste product. For archaic life on the violent young planet, the energy might have come from the Earth's inner heat and the chemical energy of minerals. The relevance of Miller's experiment has been questioned, and Oparin's idea of life emerging in a "primordial soup" has lost its popularity because of the difficulty in preventing the dissolution and hydrolysis of organic material. An alternative mechanism going back to Carl von Nägeli (1884) is prebiotic synthesis in a safer environment of adsorbed layers. In Nägeli's words, cited by Günter Wächtershäuser (2007), *probably*

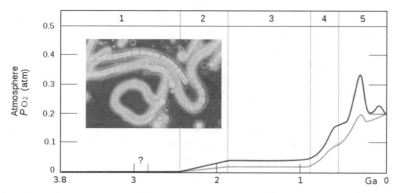

Fig. 4.5 Change in the partial pressure of atmospheric oxygen on the eon scale. The *upper red* and *lower green lines* represent the range of the estimates. *Inset*: cyanobacteria

it does not occur in a free body of water, but rather in the wetted surface layer of a fine porous substance (clay, sand), where the molecular forces of the solid, liquid and gaseous bodies cooperate. Wächtershäuser (1988) advanced this idea in a picture of the "iron–sulphur world", where a thin self-replicating layer of reactants spread on a catalytic mineral surface. This form of life needed neither sunlight nor oxygen, and was fed by energy from its surroundings.

Cyanobacteria, the first microorganisms capable of photosynthesis, were agents of great change on the Earth, enabling the emergence of new kinds of life forms getting energy by inhaling oxygen and slowly burning the organic matter produced further down the food chain. The transformation, extending for eons (Fig. 4.5), was brought to a conclusion in what is called the Oxygen Catastrophe or even the Oxygen Holocaust, exterminating archaic anaerobic life. For the first two eons or so, oxygen was absorbed by the oceans and soil, but in the end their capacity was exhausted, and the poisonous gas began accumulating in the atmosphere. Imagine our atmosphere starting to be filled by ammonia. We would not survive without gas masks, and animals lacking gas masks would be still worse off. But some microbes and plants may find the change beneficial, and evolve by learning to convert the poison into a nutrient or a fertilizer. One of the oldest surviving microorganisms, *archeae*, can use ammonia and even metal ions in contaminated waters as energy sources. Cyanobacteria capable of fixing atmospheric nitrogen may get away without oxygen, and help to turn a new catastrophe into another biospheric revolution. Catastrophe for some, new life for others.

James Lovelock (1979) observed that *the climate and the chemical properties of the Earth now and throughout its history seem always to have been optimal for life.* This prompted his romantic notion of a planet-size living entity, *Gaia*, named after the Greek goddess of the Earth, comprising the Earth's biosphere, atmosphere, oceans, and soil. The statement cited above is reminiscent of Voltaire's Pangloss, a spoof on Leibniz, who maintained, amidst all disasters, that everything was for the best in the best of all possible worlds. But if we replace "optimal" by "suitable", the deep truth of this notion is revealed in the coevolution of the planet's thermal state and chemistry – and life. Life changes the composition of the atmosphere, and the biosphere transforms, as Lovelock puts it, *in the flexible Gaian way of adapting to change and converting a murderous intruder into a powerful friend*. The environment of the young alien Earth was suitable for the earliest life, and the environment of the transformed planet became suitable for those living now.

An ardent environmentalist, Lovelock (2007) is worried about global warming, lest we must turn into thermophiles, but advocates nuclear energy as a replacement for fossil fuels. He notes that the Chernobyl exclusion zone became a paradise for wildlife, more tolerant to radiation than to human intrusions. Indeed, Gaia is versatile. The Earth may turn out to become optimal for other varieties of life. Luxurious jungles flourishing in the global greenhouse? Cyborgs? Soft robots? A sure forecast is that microbes will survive in any case. The romantic notion of Gaia – Earth as a self-regulating system tuned to supporting life may be as metaphoric as the Biblical account of Genesis, but we must care for our Earth even if she does not care for us.

4.4 Cells Are *Sine Qua Non*

A life-like form has to be separated from the environment. This is necessary even when it cannot yet be called properly alive. A chemical reactor must have an envelope preventing its contents from dissolving or uncontrollably escaping in another way, and only needs inputs and outputs to bring in fresh reactants and expel degraded waste. The essence of degradation is decreasing the *free energy* (Sect. 2.5). Engineers need to maintain a chemical reactor in a stationary state, continuously producing the desired product at the expense of free energy from the environment, and therefore a well-defined and guarded border is necessary. Compare it with border controls preventing (in some countries) immigrants from entering and (in countries of another kind) citizens from escaping: a chemical reactor needs both.

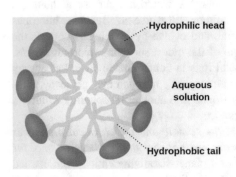

Fig. 4.6 A micelle

The same with life, starting with the grey period between chemistry and life. A modest addition to Eigen's essentials of life, function and information (Sect. 4.2), is a membrane enclosing a private space where these essentials reside. Such an enclosure could have emerged quite easily, and the idea put forward by Oparin (1924) was that cells came first, and chemical metabolism was initiated within these enclosures. Among the chaotic ingredients of the primordial organic matter, there were probably some Janus-faced chemicals, with polar heads preferring water attached to hydrocarbon chains preferring their own company or their likes. Common soap is a substance of this kind; we use it to catch fatty grime and wash it away in running water. The proper name for these molecules is *surfactants*. They tend to aggregate on the water surface, sticking their tails into the air, or to form micelles (Fig. 4.6), exposing their hydrophilic heads to water and hiding their hydrophobic tails inside.

A particular class of surfactants building up cell membranes are *lipids*. It would be natural (meaning, thermodynamically advantageous, i.e., reducing the overall free energy) for all kinds of organic admixtures in either a "soup" or an adsorbed layer to concentrate within micelles; their shell would expand, turning a primitive *membrane* into the bounding wall of the emerging chemical reactor, enclosing what would eventually become a primitive cell. It has also been shown (Hanczyc, 2003) that clay which catalyzes the synthesis of RNA also stimulates the formation of fatty acid vesicles. This links the origin of cells to the hypothesis that life originated in adsorbed layers (Sect. 4.2) rather than in a "warm pond". Some trapped chemicals may have possessed catalytic properties, and further honed them, turning them into the first protein enzymes. Both the envelope and the reactor would gradually evolve, the interior matter developing its metabolic network, and the membrane learning to be discerning, recognizing those with entry and exit visas, like a watchful border guard.

Oparin assumed that the hardware, cells and proteins, came first. He really had little idea of genes. Cells could somehow survive and perhaps even multiply by division, and only later acquire a more precise and efficient replicating apparatus. Division of fatty acid vesicles modeling primitive cells has been reproduced in the laboratory (Szostak and Zhu, 2009). The surface area of a vesicle can be increased rapidly by addition of surfactant molecules, but its volume would grow only slowly, as it is limited by the permeability of the membrane and may be counteracted osmotically, depending on the balance of solute concentrations within and outside the micelle. As the surface to volume ratio increases, the vesicle elongates, becoming mechanically unstable and breaking into smaller vesicles. Deformation to complicated shapes and division has been reproduced in the model of a vesicle bounded by a penetrable elastic membrane (Ruiz-Herrero et al, 2019). Zwicker et al (2017) came up with a model for multiplying droplets that did not even include a membrane but was based on chemical activity. An active species within the droplet degrades into "waste", which then leaves the droplet and is recycled in its environment with the help of "fuel", converting it back to the active chemical. The latter is subsequently incorporated into another droplet. If the environment becomes supersaturated, droplets lose their spherical shape, elongate, and eventually divide, whence

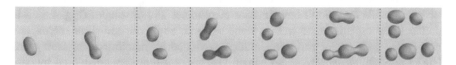

Fig. 4.7 Cycles of growth and divisions of chemically active droplets

chemical activity overcomes the coarsening tendency (Sect. 2.5) and leads instead
to cycles of growth and division (Fig. 4.7).

The great successes of genetics diverted attention from Oparin's scenario, render-
ing it all but obsolete, but it was forcefully and controversially supported by Lynn
Margulis (1970). She believed RNA to be a parasite invading an already working
cell and taking control. Her views were propped up by Freeman Dyson, an outstanding theoretical physicist and an eloquent writer. Notably, nonconformism and
love of freedom were behind this penchant. Dyson (2004) expresses this openly: *I
happen to prefer the Oparin theory, not because I think it is necessarily right but because it is unfashionable.* The rule of the "selfish gene", as it was famously dubbed
by Richard Dawkins (1976), caring only for its own perpetuity, is reminiscent of
political oppression, of invaders turning into rulers. This happened more than once
in history, like Scandinavian raiders subduing a Slavic land, "rich and and plentiful
but lacking order", known ever since as *Rus'*, or Russia, from the name of the invading band, or, moving West, acquiring and renaming Normandy before subjugating
England from this base.

Personal preferences are, of course, not proofs. Dyson (1982) developed a toy
mathematical model of an Oparin-style self-sustaining metabolic system, though it
was less elaborate and persuasive than Eigen's theory of autocatalytic cycles. Rather
naively, Dyson suggested that chemists might imitate Eigen's experiment with a
self-generated population of RNA molecules (Eigen et al, 1981), starting instead
with combinations of catalysts and metabolites in a droplet and seeing whether it arrived at a lasting homeostatic equilibrium. In parallel, he suggested an accompany-
ing simplification experiment, starting with a cell and seeing whether its metabolic
network would degrade if placed in a pampering environment. This would be a
counterpart of the experiment by Spiegelman (Mills et al, 1967) aimed at answer-
ing the question: *What will happen to the RNA molecules if the only demand made
on them is the Biblical injunction, multiply, with the biological proviso that they do
so as rapidly as possible?* It all came to "Spiegelman's Monster": a shorter chain
replicates faster, and evolving RNA got rid of one nucleotide after another, eliminating 83% of the original genome. Dyson's idea was hardly practical, and nobody
followed it; chemists and biologists hardly ever take theorists' advice, because they
know what they are able to do, and often come upon important results by sheer
luck. Oparin's scenario in its original form is unlikely; it is more plausible that dividing protocells contained RNA replicating without the help of any proteins – but
the verdict on the modern "chicken or egg" dilemma is still out.

Margulis' idea of an invader turning into a symbiont gained acceptance on another evolutionary level. The first organisms, like contemporary bacteria and archaea
(two branches of the phylogenetic tree in Fig. 4.1) were *prokaryotes*, with simply
structured cells that didn't have interior membranes and organelles (Fig. 4.8). All
multicellular organisms, as well as some bacteria, amoebae, and slime molds, are
eukaryotes with nuclei and mitochondria separated by membranes of their own.
Nuclei contain *chromosomes*, so named just because they can be colored by certain chemicals in the lab. The nucleus is the command center of the cell where the
genetic information is stored and wherefrom instructions specifying the synthesis

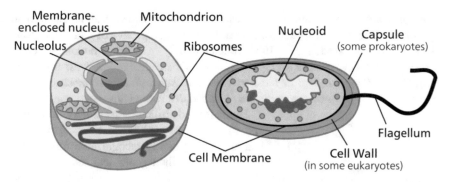

Fig. 4.8 Eukaryotic (*left*) and prokaryotic (*right*) cells (details discussed in Chap. 5)

of proteins are issued. A mitochondrion is the cell's powerhouse, producing adenosine triphosphate (ATP), the chemical currency of the cell (Sect. 5.3). The energy released when ATP loses one of its phosphates, converting into adenosine diphosphate (ADP), is used elsewhere for all the mechanical functions of the cell, as well as for driving endothermic chemical synthesis. Not unlike money, ATP takes part in many of the cell's signaling functions.

Mitochondria carry their own DNA, which prompted the hypothesis, going back to Constantin Mereschkowski (1905), that eukaryotes emerged through cell fusion, leading to the symbiosis of two dissimilar cells. This scenario was popularized by Lynn Margulis (1970). The *rationale* is that a properly functioning cell would never undertake such a *perestroika* on its own. This mechanism gained acceptance when newly developed genome sequencing techniques indicated the chimerical nature of eukaryotic cells: metabolic mitochondrial genes are similar to bacterial genomes, while informational genes in the nuclei are related to archaea. Eukaryotic cells are just archaea infected by bacteria which successfully avoided the cell's defences, but rather than killing it, settled there peacefully. This was one of the horizontal gene transfer events shown in Fig. 4.4. Another example of an organelle originating through symbiosis is the incorporation of prokaryotic cyanobacteria into plant cells as *chloroplasts* carrying out photosynthesis. Lynn Margulis (1998) believed symbiosis to be the driver of evolution's coups.

4.5 Darwin Evolved and Devolved

Once interactive mechanisms of memory and function were found, life had inadvertently evolved. Charles Darwin (1859) gestated his theory of evolution by natural selection for decades and only hastened to announce it to avoid being scooped by Alfred Russell Wallace, an outsider suddenly posting his manuscript from the far recesses of the British Empire. The idea was revolutionary at his time, and incited outrage, especially following the publication in 1871 of the book *The Descent of*

Man, provoking a caricature with Darwin's esteemed bearded head attached to an ape's body (Fig. 4.9). It looked so natural from the perspective of a hundred years since! Conrad Hal Waddington (1961) called it merely "a truism or tautology". Gunther Stent (1969) writes: *As everybody now knows, survival of the fittest is nothing but the tautology: survival of the survivors, and hence in this connection 'unfit' represents not an objective scientific but a subjective value judgement.*

Humans used artificial selection for millennia, and succeeded in evolving domesticated plants and animals to barely recognizable forms on a historical rather than geological scale. The real shock with Darwin's theory was that evolution was undirected and unintentional, defying the watchmaker analogy beloved of creationists. The reverend William Paley (admired by Darwin, while a student) contemplated, as he imagined pitching his foot against a watch rather than a stone while crossing a heath (Paley, 1809):

Fig. 4.9 Darwin's image in an 1871 caricature

When we come to inspect the watch, we perceive (what we could not discover in the stone) that its several parts are framed and put together for a purpose, e.g., that they are so formed and adjusted as to produce motion, and that motion so regulated as to point out the hour of the day; that, if the different parts had been differently shaped from what they are, of a different size from what they are, or placed after any other manner, or in any other order, than that in which they are placed, either no motion at all would have been carried on in the machine, or none which would have answered the use that is now served by it.

Indeed, Nature, the "blind watchmaker", had even equipped its creatures with internal clocks! A complicated mechanism would only survive and procreate if it "would have answered the use that is now served by it". In the formulation by Eigen (1971), *evolution at the molecular level may be considered a game in which the intelligence of the player is replaced by a selective 'instinct' for advantage among randomly occurring events* – and the game continues on the level of species. But strange as it is, the watchmaker analogy, in the guise of "irreducible complexity", persists, and not just among rednecks believing in a flat Earth. Michael Behe (2019), a biochemist dubbed the "father of Intelligent Design", argues that mutations can only destroy rather than create a complex mechanism, and the removal of any one of these mechanisms would render the entire apparatus defective.

There has been a heated polemic about whether evolution was gradual, by imperceptible steps, as Darwin supposedly thought, or through "punctuated equilibria" (Eldredge and Gould, 1972), periods of stasis interrupted by sudden conspicuous steps. Stephen Jay Gould widely publicized the idea, and it also diffused into linguistics (Dixon, 1997), where there is actually more evidence for rapid changes triggered by migrations and invasions. As for the evolution of species, it is not quite clear what time scale distinguishes "fast" from "slow". Alongside minor adaptations, there were major evolutionary transitions (Maynard Smith and Szathmáry, 1997): the emergence of eukaryotic cells, multicellular organisms, and, on a still higher level, consciousness and language. There were catastrophes causing mass extinctions that prompted movement into vacated ecological niches. It is certain that the pace of biological evolution was highly variable for different species and in different epochs but, save catastrophes, the geological record provides a clock scaled in millennia, and even on this scale it is uncertain due to scarcity of data. Even a faster human-driven evolution does not happen before our eyes; it took quite a while to turn a wolf into a chihuahua. Darwin himself just took the trouble to confront the creation myths and support Charles Lyell's views of gradual geological changes juxtaposed with the catastrophism of Georges Cuvier, while he himself recorded the fast divergence of his famous finches on the geologically young Galapagos Islands. The popularity of the theory of punctuated equilibria was attributed by critics to Gould's rhetorical skills and brilliant prose rather than to its unique message.

Eigen (1971) viewed the trajectory of the inevitable evolutionary process as an arrow defining the privileged direction of time, which applies to all living systems and is even more pronounced and meaningful than the unidirectional increase of entropy in any irreversible process. In his words, *the terms "good" and "evil" assume a meaning as soon as single information carriers start to interact and thereby mutually increase or diminish their values*. The inevitability of evolution leading to an increase of a "value function", or fitness, in whatever way it be defined, prompted creation of the various *evolutionary algorithms* (see, e.g., Ashlock, 2006). A population of programs subject to mutations is selected by their ability to solve a computational problem, with weaker ("less fit") performers eliminated on the way.

An advantage of computers over the real world is that everything goes fast. However, evolutional optimization has also been implemented *in vivo* in the experiment by Richard Lenski, started in 1988 and, not being as fast as *in silico*, continuing for many years and many thousands generations of the bacteria *E. coli*. It demonstrated the adaptation of bacteria, increasing their fitness (measured by efficiency in consuming nutrients) in different sugary solutions and at different temperatures, as well as evolving resistance to virulent phage. After more than thirty thousand generations, bacteria even managed to spontaneously acquire the ability to metabolize another organic component of the solution, citrate, which earlier generations could not utilize (Barrick et al, 2009). Neither *in vivo* nor *in silico* did an intelligent creator interfere, and it should have been sufficient to bury the watchmaker analogy forever.

Fast or slow, evolution never stops.The traditional assumption, going back to Aristotle's *Scala Naturae* (Lovejoy, 1965), is that it brings about a steady rise to higher complexity, in about the same order as in *the Great Chain of Being* (Fig. 4.10). Of course, the old drawing on the left does not imply that angelic beings in the upper row had evolved from humans (though why not? – aren't good Christians keen to rise to a heavenly paradise?), but the biblical Creation had indeed followed the path from plants to animals to man, as depicted here. A much later picture from Scientific American in the right panel depicts the same sequence, just putting man at the bottom, perhaps as a hint.

Fig. 4.10 The Great Chain of Being. *Left*: The drawing of the Great Chain of Being by Diego Valadés, 1579, up from plants to God. *Right*: A modern interpretation, down from amoeba to man

Organisms having an apparently simple morphology are naturally placed at the base of the tree of life. Innovative forms and behaviors are never abandoned by life at large, but the drive to complexity, though prevalent, is not universal. Secondary simplification is encountered at all taxonomic levels, as was already argued long ago by André Lwoff (1943), a winner of the 1965 Nobel Prize in Physiology and Medicine. An obvious example is the loss of vision by underground and deep underwater dwellers. Parasites are prone to simplification – think of social parallels. I mentioned the degradation of Spiegelman's Monster (Sect. 4.4); mutations of bacteria in Lenski's experiment also degraded some of their functions that became unnecessary in the protected laboratory environment. Rather than demolishing evolved structures, Nature often retools them for new functions, the process called *exaptation*. Whales and dolphins lost their ability to walk on land, but their front limbs evolved into flippers, which were more useful in their new environment.

Evolution does not optimize, it retains what is working in a particular environment. Optimization implies a tendency to approach a maximum or a minimum of a certain function, but there is no measure of such a vague notion as fitness, and if there were, it would have the form of a rough landscape with as many local maxima as there are biological species. Antonio Damasio (2018) uses homeostatic efficiency as such a measure, and this is also a non-quantifiable notion. Homeostasis in its traditional meaning, going back to Claude Bernard, refers to regulation keeping a system or a process within an agreeable range of parameters that allows its preservation and continuation. Any organism has this kind of a regulatory mechanism (operating until it fails at death), but in regard to particular species, it is rather a conservative force. Cockroaches have survived for three hundred million years or so without feeling Damasio's "homeostatic imperative" urging them to move on to perfection, whereas humans and humanity as a whole are bewildered by a whirlwind advance on a time scale shorter than their own lifespans that leaves no gap for social and technological homeostasis.

A species or even a larger taxonomic group may be eliminated just because of bad luck, and the biosphere as we see it certainly would not be restored in all detail if evolution was rewound and replayed. Evolution may be driven not just by strength and shrewdness, by ability to kill and avoid being killed, but by sexual preferences, from exquisite displays of paradise birds to an Indonesian monkey species whose females go for mates with large drooping noses and beer bellies. There is also a random *genetic drift*, emergence of features that neither increase nor decrease fitness, like color of eyes (before genes for blue eyes would be implanted in designer babies). Creatures may drift to either higher or lower complexity if neither gives a clear preference. Humankind also kept evolving and diversifying, sometimes by adjusting to specific environments, like high altitude, and sometimes by acquiring evolutionarily neutral features, which, however, sometimes gain significance due to cultural habits and prejudices. We are idolizing progress, and political correctness hinders discussion of long-time dysgenic consequences of some contemporary attitudes and mores – but perhaps this topic can be set aside for the time being, since humanity faces more menacing and urgent threats. Lovelock believes that we are the nervous system of Gaia, and she will not let us go down; others rely on God's good will, and I do wish they were in the right.

4.6 Lamarck's Revenge

Darwin was not the first to contemplate the ways organisms evolve without divine interference. Jean-Baptist Lamarck (1809) asserted in his first law that *in every animal [...] frequent and sustained use of any organ gradually strengthens, develops, and enlarges that organ [...]; while the constant disuse of such an organ [...] weakens it [...] until it finally disappears.* These adaptive somatic changes acquired by organisms were supposed to be passed to their offspring, thereby driving the species to perfection. Lamarck's teleological idea was noble: *le pouvoir de la vie,* striving for organisation and perfection, far more beautiful than Darwin's random genetic wandering in the dark, with its cruel extinguishing of the unfit. *To be an Error and to be Cast out is a part of God's design*, and a part of Darwin's design as well, though William Blake hardly meant it in the quoted line.

Of course, Lamarck had no idea of any mechanism for the inheritance of acquired characteristics. And Darwin himself knew nothing of genes, nor the mechanism of mutations that would become clear during the hundred years following the publication of his theory of natural selection. But while new evidence backed Darwin, Lamarck's heirloom turned out to be toxic. In Russia under Stalin, the infamous Trofim Lysenko advanced his home-grown kind of Lamarckism after cracking down upon "Mendelists–Morganists", proper geneticists basing their work on the laws of

Mendelian inheritance, confirmed and elaborated by Thomas Hunt Morgan who had by that time been awarded the 1933 Nobel Prize in Physiology or Medicine. It was a crude job: adversaries occupying themselves with flies rather than advancing Soviet agriculture were not just fired but jailed. Instead of useless model animals, Lysenko's experiments were carried out at a dedicated farm, using cows fed on chocolate in the hope that their progeny would produce more milk.

Setting these excesses aside, Darwinian and Lamarckian evolution are not always easy to distinguish. A telling example is the theory of genetic assimilation by Conrad Hal Waddington (1942). In his experiments, *Drosophila* were exposed to a toxic shock of ether, and in later experiments to a heat shock. In response to this, some flies underwent a phenotypic change (they developed a second thorax), increasing their tolerance to this environment. After selection for this phenotype for a score of generations, it was also retained under normal conditions. Waddington was inclined to interpret this in a Lamarckian way but it can be naturally interpreted as a result of genetic mutations, similar to adaptation of bacteria in Lenski's later experiments (which, however, took many more generations to develop).

Unlike firmly established genetic laws, there are no robust mechanisms for inheritance of acquired characteristics. Only gametes pass genes to offspring, while somatic cells do not inherit whatever mutations they may accumulate. August Weismann (1885) proved it by cutting tails off white mice for five generations, and seeing that tails neither disappear nor even grow shorter. However, subtle exceptions, still unclear, are now being revealed in *epigenetics*. Some phenotypic (observable, as opposed to innate) traits are due to operational details of the cells' chemical plant. They may feed back on the chemistry of chromosomes and get inherited without affecting the genetic code itself, but possibly changing the expression of genes by turning them on or off. Inherited DNA may also be modified in the maternal environment while an embryo develops.

These marginal occasions are not what I mean by Lamarck's revenge. This comes when nurture steps in to aid nature. Acquired cultural characteristics *are* inherited through custom and education by the spreading of *memes*, a word coined by Dawkins (1976) in analogy with genes. Throughout human history, memetic inheritance has led to a vast upsurge of complexity in culture and society, while the human genome has remained virtually unchanged. Yet distinguishing between genetic and memetic factors is not always easy and is often driven by value judgements. Typically, putting stress on genes is a right-wing, descending to racist, attitude, while left-wing intellectuals downplay the role of inborn features. A bizarre exception is Noam Chomsky, notoriously ultra-left on all issues, except those related to linguistics, his profession, where he has been advancing the idea of *inborn grammar* – of course, without presenting any proof about how it might be encoded.

Is the progress to complex forms irreversible? Likely yes, save for a major disaster. But human society, becoming more complex in some respects, may get simplified in others. The contemporary societal metabolism in production and distribution of both industrial and cultural goods is structured in a very complex way. There is a ladder of national and international bureaucracies, with a cornucopia of regulations, which only specialists can understand. There is a ladder of corporations, one owning the other in full or in part, with mutual relations as complex and obscure. All goods are passed to a consumer through a complicated network of distributors, agents, and promoters. Even if you see a movie, credits start with a long list of producers and sponsors, and, of course, most scientific papers end with acknowledgments to this fund or that. All this was and still is developing according to Lamarck's laws; just reinterpret the definition of an "organ" in the above quotation, and not always for the benefit of society at large, rather than owners or beneficiaries of an "organ".

On the other hand, if we look back at medieval societies, we observe that they were strictly structured in another way. Everybody was assigned to their niche, peasants to their lots, artisans to their guilds, and nobility to the ladder of vassals and suzerains. Even angels had a hierarchy of their own, from the lowest-rank privates of the heavenly army to the seraphim. Everybody's social position could be established by their dress and manner of speaking, eating, and entertaining. All this has gone. In a democracy, people are equal, and rich and poor wear jeans and sneakers in the street, even if bought at a different price. In the developed world, everybody travels, though some in business class and others in economy. There are, of course, elites, but they are not closed, and sportsmen and entertainers are on a par with royalty and financial moguls.

We can extend this picture to culture. It has known periods of sophistication and digression, and it looks like we are now in the latter phase. Compare the elaborate rhyming structure of Dante's tercets or of the classical sonnet to *vers libre*, further descending into plain strings of inexpressive words. Compare Risorgimento paintings to the contents of empty museum rooms with straight lines and squares on the walls, Baroque music to pop songs or the monotony of Philip Glass. In complete contrast, the dynamic language of modern dance exceeds in its complexity and versatility both classical ballet and folk dance. Can an organ weakened by constant disuse ever be restored? Perhaps. There are still people investing great efforts in learning to play old instruments for very little remuneration, and there are still people eager to listen to them. There are groups of painters aspiring to restore old techniques – but both money and society's attention are directed elsewhere.

We see both simplifying and elaborating tendencies at work – but the latter are closely connected to the production of goods, and this has developed into an extremely complex interconnected super-metabolic network. Politics, civil mores, and art are "parasites" feeding on this network, and they can afford simplification with no dire consequences, at least for a while. Yet they bear memes that are passed to later generations, who, being nurtured on degraded culture, may become unable to sustain the metabolic complexity essential for their survival.

Chapter 5
Cells in Motion

5.1 Why Such a Tangle?

The first open question posed by Freeman Dyson (2004) is: *Why is life so complicated?*" This is, indeed, a great inconvenience for a physicist, a habitual model builder. Look at the two networks shown in Fig. 5.1. On the left is a *model* of the metabolic network of the lowly bacterium *E. coli*, and on the right, one of the *sociograms* pioneered by Jacob Moreno (1934). There is a difference in the graphics: on the left, all arrows are directed from one protein to another one that it catalyzes, while on the right, many arrows indicating human interactions or influences are double-headed. Nevertheless, the graph on the left also contains closed autocatalytic loops. If we don't care about the meaning of the letters in the circles, we are left with an abstract graph of the kind studied by the theory of networks, which aspires to universality and does not care about meanings either. Yet, the chemical nature of any metabolic network is unique. Proteins, its nodes, have specific functions depending on their ability to recognize and bind other molecules, *ligands*. This recognition depends on specific binding sites and the specific three-dimensional configuration of the protein, and the variety of interactions among proteins far exceeds whatever we might encounter in other networks.

A network of Facebook "friends", or airline connections, or metabolism in the human body cannot be drawn on a sheet of paper. The first two can be, and certainly are, fed into a computer with a sufficiently monstrous memory, but the last one, though perhaps containing fewer elements and links, cannot be, because it is unknown. One can compile useful statistics for large networks, either abstract or specific, like the two I have just mentioned, but the statistics of a metabolic network is good only for mathematicians to play with. Here we need specifics, chemical structures, configurations of proteins, even reaction rates, which are rarely available even in the simplest cases. How many millions of working hours, how many teradollars have been invested in studies of metabolic circuits involving the production of a single protein in a single organism, often just a "model animal", like the fly *Drosophila* or the worm *C elegans*, or even the model microbe *E. coli*? There is

© Springer Nature Switzerland AG 2020
L. Pismen, *Morphogenesis Deconstructed*, The Frontiers Collection,
https://doi.org/10.1007/978-3-030-36814-2_5

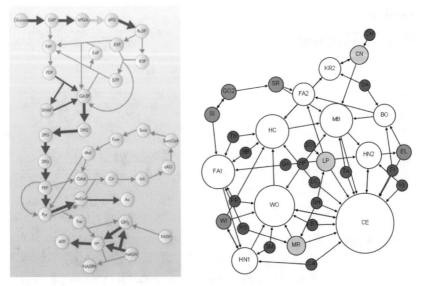

Fig. 5.1 *Left*: A metabolic network. *Right*: A social network

a proverb by Jacques Monod which says: *what is true for* E. coli, *is true for the elephant*. There is, indeed, universality up to a certain level: deep down all earthly creatures share the same chemistry, but our metabolism and our illnesses differ from those of model organisms.

Some studies of this kind may contribute to human health, and many more may bring satisfaction (and promotion) to the researcher, but we cannot set foot in the biochemical swamp, and may only observe it from a firmer ground. First of all, if such a complexity exists, it is a telling sign that it is necessary. Of course, natural selection does not arrive at the optimal design, but complexity is unavoidable because cells carry out many different tasks. There are many varieties of them in multicellular organisms, more than two hundred among thirty trillion human cells. They all carry the same genes, save occasional mistakes during replication, but express only a small fraction of them, those coding proteins which this particular cell needs to synthesize for its specific functions.

The cell structure shown in Fig. 4.8 is a rather crude caricature. A prokaryotic cell contains a *nucleotide* where most of the genetic material is concentrated, but it lacks a membrane separating the nucleus from the *cytoplasm* where the various membrane-bound *organelles* of eukaryotic cells are situated. In eukaryotes, the nucleus is further structured, including a *nucleolus*, the site where *ribosomes* are assembled. The major function of the nucleus is storage and expression of genetic material contained in chromosomes distributed within a dense mass of *chromatin*[1]. The nuclear membrane contains structural channels allowing for the passage of large molecules: proteins and RNA. Proteins have to display a distinct signal to

---

[1] Both are historic terms referring to the ease of staining, rather than to essential functions.

pass through a nuclear pore, like entering a code at a safeguarded door. The small circles in Fig. 4.8 show *ribosomes* where protein synthesis is carried out.

The *cytosol*, filling a prokaryotic cell or the cytoplasm of a eukaryotic cell outside the nucleus, is basically just salty water – but it is heavily laced with polymers that render it *viscoelastic*. The interior of a cell may look like a chaotic mess of proteins moving in all directors, sometimes carried on the shoulders of motor proteins running along "scaffolds" – actin filaments that form the *cytoskeleton* supporting the cell's mechanical integrity. Yet somehow, among all this hustle, things are getting done, and even more efficiently than they are done in a well-designed chemical factory or in a busy city. Cells are always in motion, even when they remain in place. Uri Alon (2007) expresses it poetically: *Cells are matter that dances. Structures spontaneously assemble, perform elaborate biochemical functions, and vanish effortlessly when their work is done.*

So why is the cell so complicated? You acquire complexity by a lifetime of learning, and the cell has had eons to learn, since it first began as a bag of chemicals enclosed by a lipid sheath. The complexity and efficiency of cells defies any social or industrial analogies. We still have to go a long way to bring social interactions and technologies to the level of sophistication of a bacterial cell. We can describe the way a city functions without understanding the lives of its inhabitants, busy in their various occupations. In the same way, we can try to understand how a cell works without delving into its deep intrinsic chemical mechanisms.

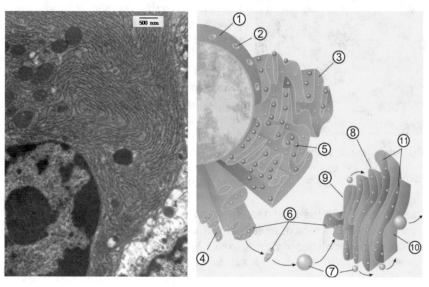

Fig. 5.2 *Left*: Micrograph of the rough endoplasmic reticulum network around the nucleus, shown in lower left-hand corner. *Small dark circles* in the network are mitochondria. *Right*: Scheme of endoplasmic reticulum and Golgi apparatus: (1) nucleus; (2) nuclear pore; (3) rough endoplasmic reticulum; (4) smooth endoplasmic reticulum; (5) ribosome; (6) transported proteins; (7) transport vesicle; (8) Golgi apparatus; (9) *cis* face of the Golgi apparatus; (10) *trans* face of the Golgi apparatus; (11) cisternae of the Golgi apparatus

5.2 Protein Traffic

The principal function of any cell is protein synthesis. In most advanced eukaryotic cells, proteins are produced on ribosomes which nestle in the maze of *rough* endoplasmic reticulum adjacent to the nucleus, wherefrom instructions are delivered by messenger RNA. A ribosome is a sophisticated molecular machine containing its own ribosomal RNA and enzymes. It scans messenger RNA molecules copied on DNA in the nucleus and carried through the cytoplasm, and assembles amino acids coded by their nucleotide sequences. RNA strands are too unstable to travel safely at elevated temperatures, which explains why there are no thermophiles among eukaryotes. The assembled proteins are packaged and distributed by *Golgi apparatus* (Fig. 5.2). The way things work is better seen in the scheme on the right than in the actual micrograph on the left, where the nucleus in the lower left corner and rough endoplasmic reticulum are easily recognizable, but further detail has to be deciphered by those in the know. *Smooth* endoplasmic reticulum is the site of less sophisticated synthesis of lipids and other chemicals.

Fig. 5.3 A kinesin molecular motor carrying protein cargo along a microtubule

When produced, the proteins are wrapped in vesicles and sent to the Golgi apparatus, usually situated in the middle of the cell. This serves as a distribution center, efficiently organized with entry gates at the *cis* side, collecting sacks called *cisternae*, and exit gates at the *trans* side, directing the product to wherever it belongs. Unwanted materials are sent to *lysosomes* where they are degraded. Vesicles are wrapped by a coating protein or *clathrin*; they bud from a membrane enclosing a sending organelle and fuse to a membrane on the receiving side, so that the transported protein never comes into contact with the watery cytosol. Both budding and fusing require recognition and expend energy.

Wherever protein traffic is directed, it would be too slow and inefficient if a clumsy polymer molecule had to diffuse through crowded cytoplasm. Instead, cells have built fast tracks, *microtubules*, and employ fast courriers – *kinesin* molecules, a kind of molecular motor carrying protein cargo (Fig. 5.3). Kinesin is a dimer, each one having a globular *head*, while its long attached strands are intertwined in a *stalk*. The dimer is structured in such a way that it literally walks upright on its track, with its two heads serving as feet. Heads, being the motor domains, are the most important parts. Just imagine having two heads and moving them exactly as you move your feet, lifting one, moving it ahead while turning around a bit, then lowering it on the track, moving the other one, and so on, all this while holding heavy cargo on the top. Where you have torso, kinesin has a stalk, and on the top, instead of a head, is a tail holding the cargo. Of course, gravity is not operational on this scale, and there is no difference between up and down, so you really can walk on your heads, provided you have at least two.

Each head has an attachment site which glues it to the track and is detached when the head is lifted. The bond between the kinesin and the microtubule must be physical rather than covalent, to facilitate attachment and detachment. Another attachment site on the tail binds it to the cargo. Whether walking on feet or on heads, you need energy. This comes from the common cellular currency – ATP. Each step involves a change of conformation, and requires one ATP molecule to be released. The cargo should be carried in a certain direction rather than wander back and forth. This is ensured by making the tracks unidirectional: the microtubules are polarized, so that most kinesin molecules walk towards their "positive" end, although some are able to switch directionality.

Some proteins are carried to the *plasma membrane* bounding the cell, to be either expelled as waste or sent out as agents of trade and communication with other cells; other proteins may come from outside as signals, nutrients, or invaders. The plasma membrane of a eukaryotic cell is organized with a sophistication matching the cell's interior (Fig. 5.4). Its basis, inherited from primitive cells, is a lipid bilayer with hydrophilic heads on both the inner and the outer sides and hydrophobic tails within. But this is only a matrix carrying built-in membrane proteins, either *integral*, imbedded in the membrane, or *peripheral*, placed at either side and easily detachable. The lipid layer is liquid, and proteins are free to move laterally within the membrane.

Small gas molecules, like oxygen and carbon dioxide (CO_2), can dissolve in the lipid layer and freely pass through the membrane, but ions and large organic molecules important for the function of the cell are tightly controlled. Movement of ions is essential, as it affects the voltage difference across the membrane. It is taken care of by *channel proteins*, commonly specialized in transporting particular ions and able to drive them *against* their concentration gradient. The channel guarded

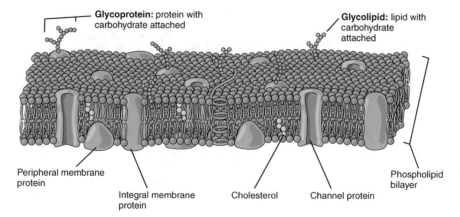

Fig. 5.4 Plasma membrane of a eukaryotic cell

by a protein can be properly called a *gate* – it may open and close in response to a voltage change or a chemical signal.

Proteins on the surface of the cell may serve as catalysts or attachment sites able to recognize and bind incoming molecules, which allows communication between the cell and the extracellular space. *Transmembrane* proteins transmit the signal to the interior of the cell by sending their own messenger there when prompted by a messenger coming from outside.

5.3 Energetics and Electrochemistry

Maintaining the cell's chemistry in a non-equilibrium state, expressing genes, assembling proteins, and moving them around, all of this costs energy. It comes in the first instance from sunlight, as plants, algae, and photosynthetic bacteria convert CO_2 and water[2] to organic matter, which goes further up the food chain as fuel for oxygen-breathing organisms.

Once gained, energy must be circulated internally to reach the right places at the right time. Energy can be compared with money: you pay for work requiring effort. Nature has not developed electronic money that can be transferred between the bank accounts of the various molecules or cellular structures, nor anything like our own symbolic paper money. Her money is material, but far more sophisticated than gold or silver coin. It is not hoarded, but transferred, used, and recreated; an animal may use about its own weight of molecular currency a day. Cellular coinage is not made up of sophisticated proteins, but small molecules that can be reversibly transformed from a high- to a low-energy state and *vice versa*.

The energy of organic fuel, either photosynthesized or ingested, is invested in the most expensive circulating coin: a nucleotide going under the abbreviation NADH. It releases a large amount of energy and acquires a positive charge turning it into

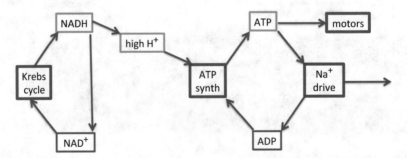

Fig. 5.5 Energy exchange in an animal cell. High energy states are in *red*, and lower energy states are in *blue*

---

[2] Both are greenhouse gases, but only CO_2, which is actually food for plants, is called "poison" by folks scared of global warming.

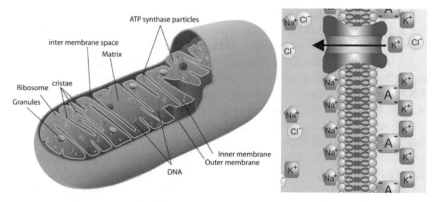

Fig. 5.6 *Left*: Section of a mitochondrion in a eukaryotic cell. *Right*: Electrochemical potential across the plasma membrane. The interior of the cell is on the right and the exterior on the left

a low-energy NAD^+ ion when it passes two hydrogen atoms to an oxygen atom to make a water molecule. NADH takes part in the *Krebs cycle*, the basic metabolic network whose discovery brought the 1937 Nobel Prize in Physiology or Medicine to Albert Szent-Györgyi and the 1953 prize to Hans Adolf Krebs and William Arthur Johnson. As it is common to all cellular processes, the purpose of this cycle, to restore NAD^+ to high energy NADH at the expense of the externally invested energy, is not attained in a straightforward way, but involves a chain of catalytic transformations, each one facilitated by a dedicated protein enzyme. The Krebs cycle is used by all aerobic organisms and leaves carbon dioxide as a waste product.

The energy stored in NADH must be further converted to other currency, to be used in a variety of cellular processes going uphill, against the thermodynamic trend. We have already encountered the most important cellular coinage, ATP, and the organelle where it is produced, the mitochondrion. The energy is released when one of its three phosphorous groups is hydrolyzed, ATP $\rightarrow$ ADP (ATriP to ADiP, P standing for phosphorus and A for adenosine, the name of the carrying nucleotide), and it is recharged when the missing phosphorus group is joined again, in the mitochondrion.

Degradation of NADH to NAD^+ is used to facilitate production of ATP, but not directly: it is first invested in the energy of a non-equilibrium distribution of hydrogen ions[3] across the mitohondrial membrane (Fig. 5.6, left). Bacterial cytosol is weakly alkaline, so the concentration of H^+ is higher outside the cell. The plasma membrane of eukaryotic cells does not support an H^+ excess, and, to create a similar environment for mitochondria, descendants of prokaryotic bacteria, the H^+ gradient is created by investing the energy stored in NADH.

---

[3] Biologists, denoting hydrogen ions by H^+, often call them "protons", but a proton, the hydrogen nucleus, can never be free. In fact, the positive ion found in aqueous solutions is a *hydronium* ion H_3O^+, a hydrogen ion attached to a water molecule. To follow the convention, we will still call this complex a "hydrogen ion", just bearing in mind that it is always hydrated.

Transport of ions through membrane channels depends both on their concentration and the voltage difference, combined in an *electrochemical potential*. Thus, to maintain equilibrium, the inside should be positively charged. But neither the bacterium nor its descendant follow this rule: they abhor equilibrium, because they need energy that is released when positive ions rush in, being driven by both concentration and voltage gradient. This energy is used by a molecular machine called ATP synthase to attach a third phosphorous group back on the ADP.

ATP can be used, among other things, to facilitate transport of ions through the plasma membrane. An important task, in particular, for nerve cells, is to maintain the balance between positive potassium K^+ and sodium Na^+ ions. Potassium ions, free to pass through their dedicated channels, are in equilibrium, which means that the concentration gradient driving them outside should be compensated by negative charge in the cytosol (Fig. 5.6, right). But then sodium ions Na^+ are driven inside both by their concentration gradient and voltage difference, in the same way as hydrogen ions rush through the mitochondrial membrane, and release energy. The protein called sodium/potassium ATPase located in the plasma membrane uses the energy stored in ATP to drive Na^+ ions back and maintain their lower concentration in the cytosol. The entire energy conversion circuit, from the Krebs cycle to maintaining the sodium/potassium balance, is sketched in Fig. 5.5.

The exchange of Na^+ and K^+ across the plasma membrane is the electrochemical basis for the propagation of an excitation impulse along an axon of a nerve cell (Hodgkin and Huxley, 1952). When the action potential is triggered (the neuron is "fired"), Na^+ channels in the membrane open and Na^+ ions rush in, reversing the polarization of the state at rest with an excess of sodium ions Na^+ on the outside and potassium ions K^+ on the inside. This stimulates neighboring Na^+ gates to open, and in this way the action potential travels down the length of the axon. In the wake of the propagating pulse, the distribution of K^+ and Na^+ is reversed and the neuron retreats locally to a refractive phase, with the potential more negative on the inside and unable to react to a new excitation. Next, K^+ channels open and the original state is restored (Fig. 5.7).

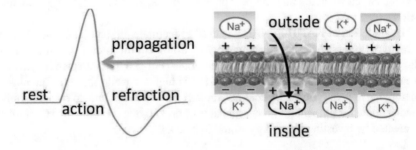

Fig. 5.7 Propagation of an impulse through an axon (see text for explanation)

5.4 Cytoskeleton

Membranes of prokaryotic cells are sturdy, and are able to protect the cell mechanically, but more sophisticated eukaryotic plasma membranes sacrifice mechanical
strength for recognition, signaling, and transport functions. The integrity of the cell
must be supported in another way, and it is the *cytoskeleton*, as is already clear from
the name, which keeps the cell together. It is also a structure built of proteins, this
time organized in a network of filaments (Fig. 5.8, left). The sturdiest filaments of
this kind, microtubules, we have already encountered in Sect. 5.2 in their capacity
as tracks for molecular motors. As the strongest structural element of the cytoskeleton, they play an important role in the process of cell division, as we'll see in the
next section. These rigid tubular structures, which hardly bend over their length, are
assembled from dimers of the *tubulin* protein, curling in a tight helix 24 nanometers
wide around a hollow center (Fig. 5.8, right). Next come intermediate filaments, 10
nanometers thick. They are the least prominent of the cytoskeletal filaments, and
differ from the other two kinds by being non-polar and more flexible.

The most numerous of all are *actin* filaments, also called *microfilaments*, assembled from actin monomers. They are only about six nanometers in diameter, and
more flexible than microtubules, but they still bend only slightly over their length.
A network of actin filaments is most dense near the plasma membrane, forming the
cell *cortex* (Fig. 5.9, left), enhancing mechanical strength where it is needed most.
The cortex is attached to a substrate or to the intercellular matrix by *focal adhesions*,
shown by green dots in this picture. In reality, in their full glory, they are complex
molecular machines composed of several distinct types of proteins and fastened by
integrin proteins crossing the plasma membrane.

What makes the cortex tough is the branching and interconnections of the actin
filaments. Special proteins nucleate their branching at attachment points (Fig. 5.9,
upper right). For steric reasons, branches are directed at 70° to the mother filaments,
but this still does not fully determine their direction, and the entire network comes
out to be quite disordered. Filaments going in the various directions come close together at some points, and they are fastened there by binding proteins. The network

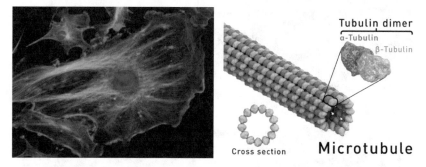

Fig. 5.8 *Left*: Eukaryotic cytoskeleton. Actin filaments are shown in *red*, microtubules in *green*,
and the nucleus in *blue*. *Right*: A microtubule

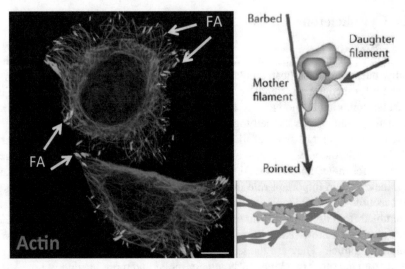

Fig. 5.9 *Left*: Actin filaments of the cell cortex (*red*) and focal adhesions (*green*). Scale bar: 10 microns. *Top right*: Branching actin filaments. The multicolored blob is the protein facilitating branching. *Bottom right*: Actin filaments tied up and stressed by myosin motors

is further cross-linked and stressed by myosin molecular motors, sometimes connected in filaments of their own and attaching their heads to nearby actin filaments (Fig. 5.9, lower right). Other proteins bundle parallel or anti-parallel filaments into *stress fibers* coming in several varieties: *ventral*, attached to focal adhesions, *dorsal*, attached at the ends to ventral ones and arching to strengthen the upper side of the cell, and transverse arcs attached to dorsal stress fibers (Fig. 5.10, left). In muscle cells, actin and myosin filaments are organized in *sarcomeric* structures, contracting due to the motor action fueled by ATP (Fig. 5.10, right).

Cytoskeletal filaments are alive, they are constantly growing, dissolving, and breaking (Fig. 5.11, left). Monomer units are attached at one end, treadmill along

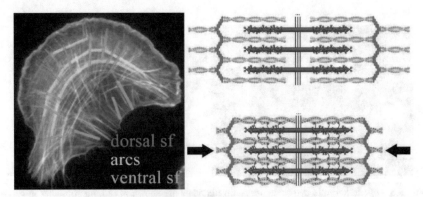

Fig. 5.10 *Left*: Stress fibers. *Right*: Relaxed and contracted sarcomere structures in the muscle

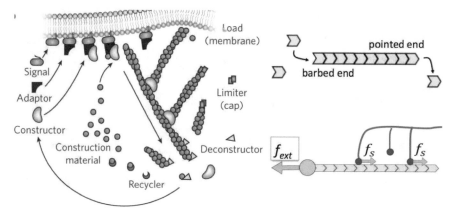

Fig. 5.11 *Left*: Recycling of an actin filament: permanent construction and deconstruction. *Top right*: Attachment and detachment of actin monomers. *Bottom right*: Stressing an actin filament to enable insertion of an actin monomer at the barbed end

the filament, and detach at the opposite end, unless protected by *capping* proteins. The schematic picture in Fig. 5.11 (top right) shows actin monomers attaching at the *barbed* end and breaking off at the *pointed* end. An entire filament can also rapidly depolymerize and shrink. Sturdy microtubules live on the average only five to ten minutes, and a network of actin filaments can "fluidize" under excessive stretch. In a filament at rest, the barbed end is attached to another filament, to a focal adhesion, or to the plasma membrane, and must be detached for a moment to enable another monomer unit to be inserted, so that the treadmill can keep moving. Stressing the filament, either by external force or with the help of attached myosin motors (Fig. 5.11, bottom right), alleviates insertion and makes the filament grow faster.

The principal cytoskeletal components, actin filaments and microtubules "crosstalk" in a number of ways (Dogterom and Koenderink, 2019). Dynamic links attach actin bundles to the plus ends of growing microtubules, thereby guiding their mutual alignment. On the other hand, the actin cortex near the plasma membrane anchors microtubules or puts a physical barrier on their growth, preventing them from hitting the plasma membrane (Fig. 5.12). Conversely, actin filaments nucleate at the ends of the microtubule.

Why does Nature prefer dynamic structures that ceaselessly rebuild themselves, while remaining stationary in the long run? Treadmilling of actin monomers costs energy, and so does fluctuating stress due to attachment and detachment of myosin motors. You can get tired even holding a hand horizontally in the air without doing any work. The benefit is flexibility, readiness to reshape and move at any moment following external inputs or internal needs. A crude analogy is the advantage of movable type, Gutenberg's invention, over woodblock printing. The cell uses amino acids to synthesize proteins, which can be broken back when their work is done to use the parts to make other proteins. On a larger scale, protein units are strung together to form filaments, and, going further up, cells are joined to tissues and com-

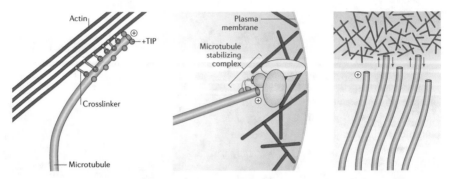

Fig. 5.12 Interaction between actin filaments and microtubules. *Left*: Microtubule guidance. *Center*: Microtubule anchoring. *Right*: Actin barrier

mit suicide by *apoptosis* when required to do so by the organism. Internal signaling and energy exchange play the role of a typesetter.

The amazing structure of the cell has been honed for eons by unicellular organisms; for them, it is a matter of survival. Animals and plants may not need it, at least not in all their cells and not all the time – but Nature is conservative, she does not abandon solutions once they are found to work. Nature cares very much about this structure, dedicating to it a substantial fraction of genomes, highly conserved along the evolutionary tree; it is estimated that between 2 and 5% of the *Homo sapiens* genome codes proteins taking part in the assembly of actin filaments alone. Researchers care no less about it, working to reveal minute details from the molecular to the mechanical level. Besides a plethora of papers, there are quite a few monographs, e.g., Howard (2001), Lee (2013).

5.5 Cells Divide

Not only the matter within cells but cells themselves are transient and expendable. We generate a complete new set of 30 trillion cells every 2 weeks, and their proliferation is balanced by death. Not all cells multiply and die, brain cells have to survive to keep our memories in their circuits, but some cells of a multicellular organism may be damaged, as our skin cells are all the time, or become superfluous and be issued a command to commit suicide, called *apoptosis*, or programmed cell death. It is more than it's worth for an organism to care for a sick cell; and neither do animals care for their sick. Only humans do it, and only at a certain stage of developed civilization which may not last forever.

Proliferation of cells is essential, as is proliferation of a species, and cell division is involved in both through its two kinds: *mitosis* and *meiosis*. Mitosis is the normal procedure for division of somatic cells. It requires duplication of all its components: genome, organelles, cytoskeleton, and cytosol. The most important part is duplicating the genome. Between cell divisions, in what is called the *interphase*, chromosomes are not seen distinctly in the nucleus (Fig. 5.13, upper left), but when it comes to division, access to the genome is needed. The nucleolus disperses, and chromosomes become distinctly visible on a micrograph. Each chromosome is located in its own region, which prevents mismatching and entanglement at the next stage. The strands of the double helix are further opened up with the help of a special protein and duplicated to form a pair of identical sister *chromatids* glued together (Fig. 5.13, upper right). They need to stay cheek by jowl at this stage, so that each daughter cell can get a complete set of chromosomes, rather than a random collection where some would be missing and others doubled.

The mechanics of mitosis is facilitated in animal cells by the *centrosome*, an organelle that has evolved relatively late and is missing in plants and fungi. The centrosome consists of two *centrioles*. When the cell starts to divide, microtubules are nucleated at the centrioles, and, as they elongate, push the centrioles apart, forming a *mitotic spindle* (Fig. 5.13, upper right) which extends to move the centrioles to opposite poles of the cell. Chromatids, set free by a dissolving nuclear membrane, attach to the strings of the spindle, and the two sisters are pulled apart (Fig. 5.13, lower left), concentrating in the two hemispheres. A cleavage develops between them, and an actomyosin ring formed in the equatorial plane constricts to pull the

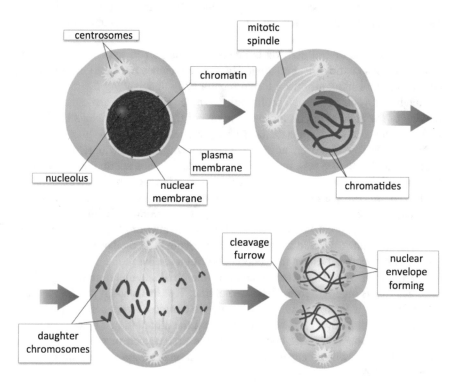

Fig. 5.13 Mitosis of an animal cell (see text for explanation)

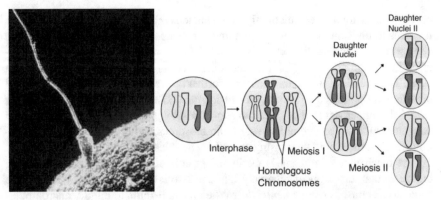

Fig. 5.14 *Left*: Sperm and egg fusing. *Right*: Meiosis. Paternal and maternal chromosomes are distinguished by color

daughter cells apart (Fig. 5.13, lower right). Once separated, each cell restores the interphase structure seen in the upper left panel of the picture.

A succinct description of this wonderful phenomenon avoids many details. Not only the genome but all organelles have to duplicate to create properly functioning cells, and mistakes are mortal – not for the organism as a whole, but for the newborn cell.

A special kind of cell division is meiosis, taking part in sex cells, or *gametes*, rather than in somatic cells. Paradoxically, the term comes from the Greek word for decreasing, lessening, also used for an intentionally understating figure of speech, although it is a stage of sexual proliferation. The origin of the term comes from the halving of the number of chromosomes in one of the two cell divisions involved in meiosis. The need for this was realized after some wanderings by August Weismann (Churchill, 2010). However, no cell comes out with a half-genome, because the process starts with two sets, maternal and paternal, when a sperm enters the ovum (Fig. 5.14, left) to form a *diploid zygote* with a double set of chromosomes.

The sperm cell, the father's contribution, is small and cheap, little more than a nucleus with the proud genome and a centriole taken along to facilitate impending cell divisions, equipped with a *flagellum* for fast locomotion, apparently, needed to win over the competition in its run toward the ovum. The mother's ovum is much larger, as it contains, besides the genome, nutrition for the future offspring and the necessary organelles. The latter include mitochondria, and therefore the mitochondrial genome is inherited from the maternal line. This has made it possible to trace the matrilineal genealogy of all living humans to our most recent common ancestor, the "mitochondrial Eve". An estimate based on the mutation rate places her about 150 thousand years ago – of course not as the first woman of the species *Homo Sapiens*, but somewhere between branching from other *Homo* species lines and the spread out of Africa.

The zygote first doubles its genome, while keeping each pair of homologous (sister) chromosomes bound together (Fig. 5.14, right). Next comes the first meiotic division, which, unlike mitosis, does not involve duplicating the genome. The daughter cells keep each parent's homologous pair, and genetic material exchanges between them through crossing over (Sect. 4.2). The last meiotic division directs each chromosome from homologous pairs to a separate daughter cell, leaving four cells with the standard set of chromosomes.

5.6 Crawling and Swimming

Unicellular organisms need to move. This is not an easy task when traveling on a solid substrate. Focal adhesions cannot step like feet, they have to be released at the hind end and created anew ahead, and motion requires realigning the entire structure of the cytoskeleton. Although migration is more relevant for the lives of microbes, crawling motion is most often studied using isolated cultured animal cells and even cell fragments lacking a nucleus. Why? It is a good way to explore the mechanics of an eukaryotic cell and to approach the far more difficult problem of cell migration within a multicellular organism, relevant for development, wound healing, and the spread of cancer.

Why would a cell move? It can be driven by a chemical gradient, which is certainly important for a microbe looking for nutrition – this is *chemotaxis*. It can also be driven by a gradient in properties of the substrate; the cell prefers it to be rigid because it must exert forces on the substrate to push itself forward, so it moves to harder ground by *durotaxis* (Schwarz and Safran, 2013). The direction of motion can also be determined by the shape of the cell itself, its *polarization*. This brings about an element of inertia: when the driving impulse disappears, the cell will continue to move in the same direction. This may be responsible for "run-and-tumble" motion: a microbe moves ahead, stops for a while, realigns, and changes direction – not a bad strategy for exploring the neighborhood in search for food (more on this in Sect. 7.3).

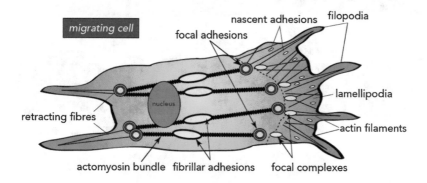

Fig. 5.15 A crawling cell

There are different ways to crawl. The leading edge is often an almost two-dimensional protrusion of the actin mesh, a "thin-sheet foot", translated into biologists' Latin as *lamellipodium*. It may inch ahead by tentatively protruding *filopodia* – another foot in the term, though they are not feet at all but aids to explore the way ahead: they can be either withdrawn or strengthened by actin filaments to pull the leading edge ahead (Fig. 5.15). Nascent focal adhesions are formed at the lamellipodium and mature as the bulk of the cell advances. The actin network depolymerizes at the back end, and actin monomers are transported to the front to polymerize there. This is a slow process limiting the speed to about the cell length per minute.

Otherwise, protrusions may be resisted by the tension in the plasma membrane, keeping the leading edge smooth and the shape elongated normally to the direction of motion (Fig. 5.16, top left). This is characteristic of the shape of *keratocytes* taken from fish skin (Ben-Zvi et al, 2008), beloved by experimentalists because of their relatively fast and persistently directed motion. A completely different way of moving, independent of adhesion to the substrate, is *blebbing*, extending spherical protrusions (Fig. 5.16, top right), which is more characteristic of motion in the three-dimensional environment of tissues and sometimes precludes apoptosis, cell death. The lower panels of Fig. 5.16 show a map of the stresses in the substrate. These are highly concentrated in adhesion-dependent motion but weak and evenly distributed in motion by blebbing (Bergert et al, 2015).

A peculiar manner of motion is used by the bacterium *Listeria* to propel itself through the cytoplasm of an infected cell by constructing behind it a tail made from

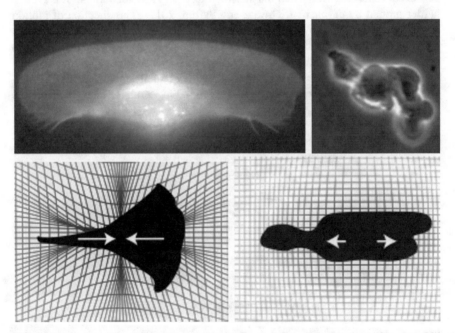

Fig. 5.16 *Left*: A crawling fish keratocyte. The shading is lighter at the thin front edge; the white spot is the nucleus. *Right*: Blebbing motion

   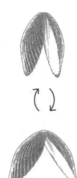

Fig. 5.17 *Left*: Demonstration of the reversibility of Stokes flow. *Right*: A reversible change of configuration

a cross-linked actin network. It is pushed ahead as actin monomer units squeeze in to join growing filaments at their barbed ends attached to the bacterium (Mogilner and Oster, 1996), in the same way as filaments attached to the plasma membrane grow within a cell (Sect. 5.4).

Swimming is the most natural form of locomotion for microbes. As they are small, inertia plays no role in their motion. Its effect is measured by the Reynolds number: size times velocity divided by kinematic viscosity. With a size of about 10 microns, a microbe would have to swim as fast as a torpedo to bring the Reynolds number close to unity. Viscous (Stokes) flow at low Reynolds number is more amenable to mathematical analysis; the classic book on swimming organisms by James Lighthill (1975) has not aged, but quite a lot has been added since then (Lauga, 2016).

Stokes flow is dissipative but, paradoxically, it is reversible in some respects. A narrow colored strip in a narrow gap between two cylinders spreads out into a form-less cloud when the inner cylinder slowly rotates, but gathers back almost precisely (just a bit blurred by diffusion) when the direction of rotation is reversed (Fig. 5.17, three panels on the left). Therefore a swimmer changing its configuration in a re-versible way, as a clam would do (Fig. 5.17, right), cannot advance, by the *scallop theorem* due to Purcell (1977).

Microscopic swimmers avoid this injunction by deforming their bodies and at-tached utensils in a variety of fancy ways. The most common propulsion aid is a *flagellum* ("whip" in Latin), a thin helical hollow tube. Bacteria developed a real motor, with stators driving it, using energy from either sodium or hydrogen ion flux through the membrane (Fig. 5.18, left). Rotation of the helix is balanced by the body rotation in the opposite direction. Eukaryotic flagella do not rotate, but they can beat in various fashions (Fig. 5.18, right). Sperm propels itself in this way.

Some organisms have several flagella, and some have their entire body covered with shorter hair-like *cilia* and propel themselves by waving them in a coordinated way. It is possible to swim even without external appendages, and even without

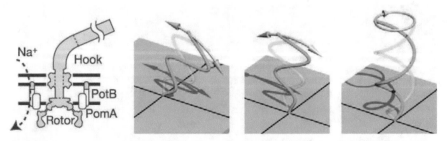

Fig. 5.18 *Left*: Bacterial flagellum motor (PomA and PotB are stators). *Right*: Flagellum beating modes

changing the shape of the body, just by inducing tangential flow in the membrane. The result is not unlike swimming with the help of cilia: waving a thin hairy cover has about the same effect as moving the surface itself tangentially. This kind of swimmer is called a *squirmer*, further classified as a *pusher* if the surface displacement in the direction of motion is stronger behind, or a *puller* if it is strongest at the forward side, or neutral if neither side prevails, as shown in the upper panels of Fig. 5.19.

Deforming or shifting a swimmer's surface tangentially does not in itself move the swimmer, but generates fluid flow, which, in its turn, moves the swimmer ahead. The lower panels of Fig. 5.19 show the flow pattern near spherical squirmers of

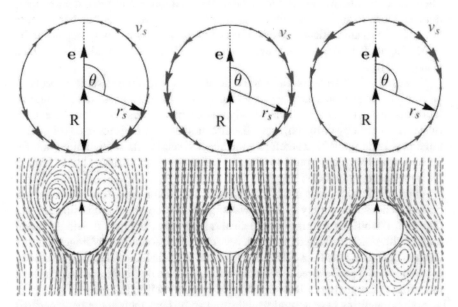

Fig. 5.19 *Top*: Tangential surface displacement by the three kinds of squirmers (pushers on the left, pullers on the right, and neutral in the center). *Lower*: The corresponding velocity fields (*red*) and stream lines (*blue*) in the frame moving with the swimmer

the three kinds (Zöttl and Stark, 2018), with paired vortices forming ahead of a pusher or behind a puller. Flow patterns are far more complicated in the case of rotating and/or beating flagella. Flow generated by a swimmer also affects others in the vicinity, creating the effect of collective motion. Flow also depends on the environment, and efficient microswimmers adjust the way they deform their bodies near surfaces or in a confined geometry. Even a scallop-like reversible swimmer can advance when swimming near a deformable surface (Trouilloud et al, 2008) or in a viscoelastic liquid (Qiu et al, 2014).

5.7 Creative Destruction

The term in the title of this section, going back to Karl Marx and Friedrich Nietzsche and popularized by Joseph Schumpeter as the essence of the capitalist economy, could have been invented by Darwin, since this is what Nature is doing, creating new species only to extinguish them and create new versions. This is also characteristic of processes within a living cell. It may have led once to innovations in the development of intracellular protein exchange machinery, as it was driving the evolution of species in a way similar to the evolution of industrial tools and social relations. Within cells, creation and destruction are routine, they are a part of all ongoing processes, as we noted when contemplating the ever dissolving and renewing cytoskeleton (Sect. 5.4). But Alon's allegory of dancing matter (Sect. 5.1) is too light-hearted. Much effort is needed to sustain this dance (well, learning to dance is not effortless either).

A protein can do its assigned job only if it is properly folded in a certain conformation. Conformations, generally, depend on the sequence of amino acids but not in a unique way. The classic assumption (Anfinsen, 1973) is the "thermodynamic hypothesis": the conformation should correspond to the overall energy minimum. However, this does not always work. Often, there are several stable minima with different energies, and the lowest one may be hard to attain, so the protein will more readily fold to a thermodynamically metastable configuration and stay there. Catalytic action usually depends on certain ordered structures within a properly folded form, and it may happen that it is a metastable state that contains such structures and therefore has a functional role.

A folded form is sustained by physical forces rather than covalent bonds forming the polymer chain. The major factors are, first, interactions with the medium, which cause the protein to fold in order to isolate its hydrophobic groups from the aqueous cytosol, and, second, hydrogen bonds formed by sharing a hydrogen atom between polar groups within the protein chain, alongside weaker van der Waals (dipole) interactions. Raising the temperature disrupts all such kinds of bonds (see Sect. 1.4), and therefore life is restricted to a rather narrow temperature interval: only a few degrees Celsius separate hypothermia from fever in humans. The catalytic activity grows with temperature but drops sharply beyond the maximum. Thermophilic bac-

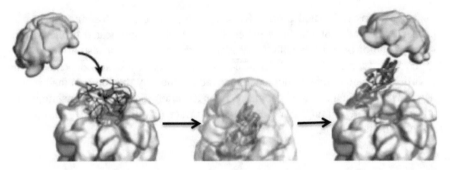

Fig. 5.20 Binding, encapsulation, refolding, and release of a misfolded protein by a chaperone

teria somehow manage to shift the maximum of their activity to 80–85°C but it is still unknown how they stabilize their proteins.

A problem arises if a sequence of amino acids properly assembled into a protein, as specified by genes, *misfolds*. Any efficiently organized process should take care of such emergencies, and Nature does. The *proteome*, as the entirety of a cell's proteins is called, includes special guards known as *chaperones*, named after old maids who in the prudish olden days accompanied society girls to keep them out of trouble. Cellular, like human, chaperones are often called for (expressed) in response to more troublesome circumstances, like elevated temperatures or overcrowding which could hamper proper folding, and help other proteins to attain their native conformation. When trouble occurs, a chaperone unfolds a misfolded protein, tries to refold it in the correct fashion, and, if it fails, depolymerizes it (an extreme action a human chaperone should never do, although "honor killings" by family members happen in some communities even today). Large chaperones can even enclose a misfolded protein in a kind of cage, protecting it from disturbing surroundings, refold it, and release it, as shown in Fig. 5.20.

It takes no less effort to degrade proteins when they turn toxic or have just done their job, than to shape them properly. Abnormalities in this process may lead to serious diseases, as misfolded or damaged proteins are potentially harmful; they are vulnerable to oxidative damage, and accumulate with age. Creative destruction on the nanoscale turns out to be not just a spontaneous event but a tightly controlled action, and studies of protein degradation systems have been honored by three Nobel prizes, first, to Christian de Duve in 1974 and again in 2004 and 2016. The major demolition machine is a protein complex called *proteasome* guided by *ubiquitin*, a regulatory protein so named because it occurs *ubiquitously* in most tissues of eukaryotic organisms. Its role is to mark proteins for degradation by proteasome, to avoid damaging useful proteins. The 2004 Nobel Prize in Chemistry was awarded for the discovery of this mechanism (Hershko and Ciechanover, 1998). The selective degradation of damaged proteins enables cells to limit the extent of oxidative damage and minimize the dangers of aging and diseases.

Chapter 6
Cells United

6.1 Why Develop?

Why should multicellular organisms exist at all? Life managed without them for three eons. Lynn Margulis (1997) rated most highly bacteria that "can do everything but talk", and with apparent approval quotes Nietzsche: *"the Earth is a beautiful place, but it has a pox called man"*. There is a sort of self-hatred tendency among radical intellectuals: they blame everything on the tribe they belong to, be it, on different levels of discussion, animals, the species *Homo*, or the West.

Perhaps life indeed could do without what is commonly called its higher forms, as it did for three eons, but even among unicellular organisms there are species agreeing to forgo their individuality. Bacteria crowd into *biofilms*, attaching to a surface and to each other, which may protect them in a harsh environment. Not unlike cells in a large organism, they are embedded in an extracellular matrix and communicate by "quorum testing" (more on this in Sect. 7.3), which allows them to coordinate gene expression depending on the local population density.

Plasmodial slime mold joins many nuclei into a huge cell, developing into a tubular network spreading out up to a meter long (Fig. 6.1). It can be a clever planner: if put in a maze, the network reshapes to connect two exits by the shortest path. If pieces of food are put at certain locations, slime mold connects them by an optimal road network taking into account density of traffic between nodes. This feature comes out naturally because tubes thicken when the flux of nutrients between the nodes increases. The planning

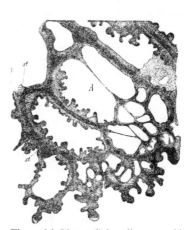

Fig. 6.1 Plasmodial slime mold (Haeckel, 1878)

© Springer Nature Switzerland AG 2020
L. Pismen, *Morphogenesis Deconstructed*, The Frontiers Collection,
https://doi.org/10.1007/978-3-030-36814-2_6

abilities of slime mold were tested in a harder task: designing a railroad network in the Tokyo metropolitan area (Tero et al, 2010).

Cellular slime mold gathers in colonies with sophisticated shapes and starts moving as a single body when food is in short supply. It is a natural step from there to a tentative diversification of the roles played by different cells and the emergence of specialists expressing and employing different kinds of proteins. Slime molds take the first step in this direction: when the colony is starving, some cells become infertile, forming a stalk supporting the remaining cells, most of which can bring viable spores. This kind of division of labor came to its extreme in extant, truly multicellular organisms: somatic cells do all the hard work to sustain gametes, which alone pass their genes to future generations.

Evolution to multicellularity by cells gathering together or the development of internal membranes in a multinuclear cell appears to be easy (Fig. 6.2). It brings advantages of increasing size and division of labor, important both in the "arms race" and in the quest for economic efficiency. It is estimated that it must have occurred many times, even in prokaryotes, but in most cases led to dead-end lineages. Extinctions might have happened because a union of cells can be as problematic as a society of selfish people. An occasional mutant may place the well-being of its own offsprings above the well-being of the organism as a whole, as happens with human dynasties. Cooperation among cells, supported by stringent laws and watchfully policed, should develop to suppress such tendencies. In the organisms that have survived to our day, all cells carry the same genetic information; all of them are offspring of the four identical cells formed by meiosis (Sect. 5.5). Yet, even now, egotistic mutants rebel: this is cancer, multiplying with no concerns but able only to kill the affected organism and die itself leaving no progeny.

Another scenario, advocated by Lynn Margulis (1998) as a driver of all crucial stages of evolution, is symbiosis, leading to mergers of genetic lines when both species become unable to survive alone. Mereschkowski (1905) coined the term "symbiogenesis" as the origin of evolutionary novelty via symbiosis. Symbiosis often brings about evolutionary advantages: different species with their specific genes differ in their capabilities, and pooling them together helps them both to survive and prosper. The contrast between symbiosis and Darwinian struggle for survival was even perceived in social terms, as a contrast between socialist and capitalist attitudes; Margulis (1997) cites prince Pyotr Kropotkin, the 19th century Russian

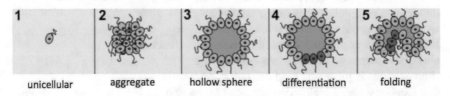

Fig. 6.2 Haeckel's hypothesis: evolution of a colony of cells into a multicellular organism (Haeckel, 1874)

aristocrat turned anarchist, juxtaposing capitalist competition with a utopian society of "mutual aid".

Symbiosis is ubiquitous in all walks of life, including consortia of flagellated and phototrophic bacteria (think of an airplane as a union of a motor and wings); lichen, a composite of algae or cyanobacteria incorporated in fungal tissues; and deep-water fish carrying luminous bacteria. It does not necessarily benefit both symbionts and does not necessarily lead them to join up into a single organism. A multitude of bacteria, far more numerous than our own cells, are our symbionts, helping us in many tasks, but this does not mean that our *ego* is partly human and partly bacterial (although Margulis might have said it is). The formation of *chimeras* combining cells with different genomes might have played a role in the emergence of multicellular organisms.

The first truly developed multicellular organisms in the fossil record are *Ediacarans*, leafy creatures growing up to several meters tall, that could even have a sort of fractal structure of fronds within fronds (Fig. 6.3). They were first classified as animals, although they were sessile and had neither mouth nor guts nor anus nor limbs, and were more like fungi or rather a separate taxa altogether. They flourished about six hundred million years ago, holding onto the sea floor, but their line became an evolutionary dead end.

Ediacarans became extinct prior to the "Cambrian explosion", the evolutionary miracle which produced, in a geologically short time around 540 million years ago, all the basic body plans of modern animals, alongside less fortunate experiments with exotic forms. The diversity of animal species further expanded in the Ordovician Biodiversification Event starting 490 million years ago. Newly evolving animals learned to move, or burrow into the sea floor, and some of them became predators, initiating the arms race between prey and predator, with faster motion and sharper senses essential to both. There would be no more "mutual aid". This was a runaway tooth-and-claw struggle for survival.

What caused this burst of innovation? It may have been a sharp increase in the oxygen concentration (see Fig. 4.5). Besides enabling large animals to breathe, it made possible the formation of the ozone (O_3) layer shielding life from lethal ultraviolet radiation. Massive global warming and the spread of shallow seas rich in minerals brought by erosion (everything we most dread today) may have contributed to this bold experiment. Some skeptics say that the Cambrian explosion was actually an explosion of fossils, as similar forms may have existed earlier but were too small and soft to leave tangible remains. However, the "snowball Earth" period preceding the Ediacaran would have been a tough time to innovate.

Fig. 6.3 Ediacaran fossils (McMenamin, 2018). *Scale bar* on the right in cm

6.2 Body Plan

The emergence of animals with elaborate body plans was the crucial result of the Cambrian explosion. The main structural features were bilateral symmetry and layering of tissues, common to phyla[1] emerging in the Cambrian explosion. Brainless *Cnidaria*, including jellyfish, corals, and polyps, having a different, centrally symmetric arrangement without layering, as well as still more primitive sponges, probably existed before the Cambrian, though this has not yet been verified by an undisputed fossil record. The most successful phyla with the novel body plan were *arthropods* (including insects, spiders, and crustaceans), *molluscs* (including not just snails but the highly intelligent octopuses and squid), and *chordates* (ranging from extinct creatures with a stiff rod of cartilage inside and no skull through fish to humans). The grading here is done according to the registered number of species, with the "highest" chordates, supported by an internal skeleton, just in third place, above worms but below molluscs and arthropods (Fig. 6.4, left), which count over a million species, dwarfing all the rest taken together.

Many kinds of animals not fitting into any surviving phyla, and exotic creatures, such as the five-eyed *Opabinia*, were eliminated in the course of the Darwinian struggle for survival. Stephen Jay Gould (1989) vividly describes the drama of identifying mineralized remains of soft-bodied creatures, which died in a mud slide in shallow Cambrian waters and were mislabeled when first discovered, as the sole representatives of extinct phyla with peculiar forms. We can only guess whether inferior skills and anatomy or bad luck are to blame for their extinction.

Diversity is likely to burst upon the scene when a new range of possibilities opens, as happened when multicellular life emerged, but to decrease in further stages of evolution. Take something which happened very recently, for example. There were many operating systems before Microsoft Windows and macOS superseded all the rest for common folks, with Linux preferred by professionals. There were many word processors, and only Microsoft Word remained to reign supreme, with LaTeX surviving in a special niche. There are substantial differences between extant forms

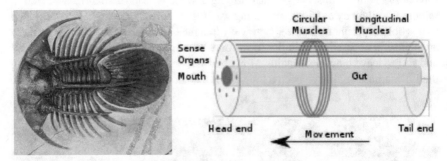

Fig. 6.4 *Left*: A trilobite (a fossil arthropod). *Right*: The bilateral triploblastic body plan

---

[1] *Phylum*, the taxonomic rank below the Linnaean *kingdom* and above *class*, was introduced by Haeckel.

that could have been intermixed originally. As Stephen Jay Gould (1989) formulates it, *the removal of most groups by extinction leaves large morphological gaps among the survivors*. Further diversification can be triggered by environmental innovations, like the emergence of new genera after great extinctions or new operating systems when smartphones burst in.

The three tissue layers of the new kind of *triploblastic* animals are already formed in the embryo. The outer layer forms skin, while the middle one forms muscles and all internal organs, except the digestive tract, which develops from the inner layer. This body plan is absent in primitive animals originating prior to the Cambrian, like sponges and medusas. In most cases, it includes a body cavity (*coelom*) containing the internal organs. Another innovative feature is bilateral symmetry, distinguishing between the top and bottom sides, and distinct front and back ends (Fig. 6.4, right).

This arrangement looks to us very natural, perhaps because we are like this ourselves, and so are our cars, and even our dwellings, both family homes and palaces, with the main entrance and large windows at the front and garbage removed at the back. Both in cars and in homes, the external walls and the interior are built from different materials and decorated in a different way, and they hold us, as their essential internal organs, in their cavities. Of course, when moving, it is good to look ahead, but this is not obviously necessary for houses, and, indeed, sophisticated architects may design centrally symmetric forms. Echinoderms (e.g., starfish) abandoned the bilateral symmetry of their ancestors after adopting a sedentary lifestyle – but it was retained by their larvae.

Here we come to something that appears strange and even wasteful. Our forms develop directly from the embryo gradually forming in the womb. Recognizably human features do not come in right away. In the early 19th century, Johann Friedrich Meckel and Étienne Serres drew attention to the similarity between all early embryonal forms, repeating the course of evolution. Ernst Haeckel (1874) summed it up in the phrase "ontogeny recapitulates phylogeny", as illustrated by Fig. 6.5. But the development of many animals is not as straightforward.

Earlier forms can be observed directly in animals hatched from laid eggs. Even among vertebrates, tadpoles look and live like fish before turning into adult amphibians. But the metamorphosis from a juvenile to an adult form can be still more radical. The novelist Anatole France once observed that he would design humans as butterflies, so that they would first work as sexless wormlike caterpillars and then, after acquiring the necessary means, enjoy freedom and love – not like us, suffering first from efforts and anxieties and then from infirmities and sickness. But would you enjoy the state of a pupa, suspended lifeless while gradually shedding the forms of your youth and re-emerging as a strange colorful creature?

Many echinoderm and arthropod species go through a larval stage, and some have even two or more distinct larval phases. This is the case of *indirect development*: a package of cells that are destined to grow into an adult form is set aside, while the embryo develops into a larva with a different body plan. Freeman Dyson (2004) describes the advantages of this arrangement, not unlike the vision of Anatole France: *the embryo provides life support to the adult, [. . .] whereas the adult is free to evolve elaborate and fine-tuned structures*. Davidson et al (1995) argued that

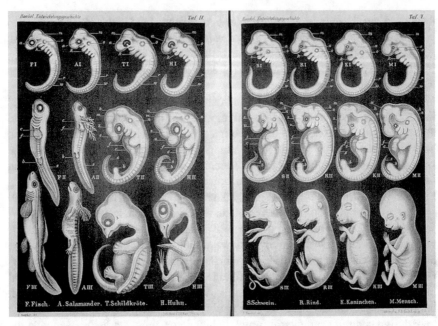

Fig. 6.5 Stages in the development of embryos of (*left to right*) fish, salamander, turtle, chick, pig, cow, rabbit, and human, illustrating the thesis put forward by Haeckel (1874)

indirect development was the original choice in all animal phyla, because it is shared by cnidarians and bilaterians, and therefore must have appeared before their divergence. This argument is not decisive, since the shared mode of development can be a result of convergence at a later stage of evolution. Donald Williamson (1992) controversially attributed indirect development to hybridization – and, of course, he was supported by Lynn Margulis, the champion of symbiosis, who had written the introduction to his book. According to Williamson, Cambrian animals were *chimeras* carrying genes of different taxa, so that the development of larvae and adults was driven by different genes, and it was this genetic interchange that was driving the radiation of new forms.

A lively field of biological research, evolutionary developmental biology, nicknamed *evo-devo*, complete with its special journals, reviews, and meetings, is dedicated to solving these and related riddles. It is driven by accumulating genetic, regulatory, and ecological evidence, but more questions arise than are solved on the way. What appeared so neatly ordered to Haeckel, is now placed under a question mark. Even the concept of phyla as a useful classification concept is doubted (Dunn and Hejnol, 2016), in view of affinities and convergences between different genetic lines.

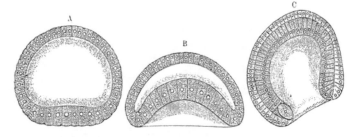

Fig. 6.6 Gastrulation: (A) blastula, (B) invagination, (C) gastrula

6.3 Signals and Patterns

Whether direct or indirect, development of an animal from an egg appears to be miraculous, far beyond anything our most advanced technologies can offer or even imagine. It starts in an innocuous way, with the initial four cells formed at meiosis (Sect. 5.5) dividing to form a shell of undifferentiated cells, called the *blastula*, surrounding the yolk. Then something special happens: *gastrulation – truly the most important time in your life*, as Lewis Wolpert quipped. It starts with invagination breaking the spherical symmetry of the blastula and turning it into the *gastrula* with distinct inner and outer layers, the *endoderm* and *ectoderm* (Fig. 6.6). This is followed by differentiation of the *mesoderm* to form the three layers of a triploblastic animal that have to develop further as directed by the genetic program.

Beyond the general body plan, every detail has to be made according to specifications and put in its proper place. All this requires communication among cells, some kind of signaling emanating from a certain source and controlled genetically. The role of such organizing centers in defining a principal body axis was already understood in the early 20th century. Ethel Browne (1909) observed that cells near the oral tip of a hydra[2] induce a secondary body axis in another hydra when transplanted into its body column. This work preceded the celebrated discovery of the *Spemann organizer*, defining the dorsal–ventral (up/down) axis in frogs (Spemann and Mangold, 1924), which earnt Hans Spemann the 1935 Nobel Prize in Physiology or Medicine[3].

Once there is an axis and an organizing center, it is natural to expect some gradients along this direction. Asymmetry between the front and back ends, and between the dorsal and ventral sides, indicates that cells must be polarized in a certain way. From about 1915, Charles Manning Child (1941) studied metabolic gradients, e.g., of oxygen, but this line of work was superseded by genetic studies. Biologists never took seriously the symmetry-breaking mechanism proposed by Turing (1952)

---

[2] A primitive marine polyp, which attracted the attention of biologists because of its outstanding ability to regenerate, even from a disorganized clump of cells.

[3] Hilde Mangold was Spemann's PhD student. She died by accident in 1926, but Ethel Browne might have shared the prize if the attitude to women scientists had been different at the time.

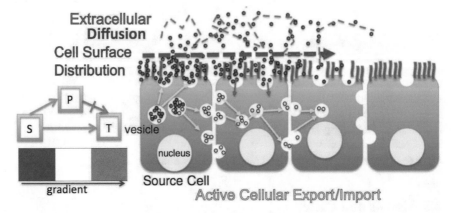

Fig. 6.7 Possible ways to transmit signals from a source cell. A scheme of the feed-forward motif and Wolpert's "French flag" are shown in the *lower left*

(which does not include polarization), but biophysicists were, and still are, trying to devise general rational schemes of development driven by chemical signaling. Hans Meinhardt (1982) writes that *it turned out that interactions employing relatively few components are able to describe elementary steps in surprising detail.* All morphogenetic patterns are dissipative structures in a broad sense, as they are actively driven and sustained far from equilibrium – but the physicist's notion of a detailed description is quite different from that of a biologist aiming to find out which particular protein is doing a particular job in a particular place, and how and why it is expressed there.

If there is a firm general principle that can be traced to Turing, it is that signaling and interaction schemes should include both activation and inhibition. The basic component of any such scheme is the *feed-forward* motif, $S \to P$, $S \to T$, $P \dashv T$, which includes two activating ($\to$) links with different thresholds initiated by the same signal S (induced by a morphogen), and an inhibiting ($\dashv$) link from the intermediate protein P to the target. This scheme generates the classic "French flag" pattern (Wolpert, 1969) shown in the lower left part of Fig. 6.7, with the target T expressed in the central (white) interval, where the signal level is below the higher threshold of the link to the protein P and above the lower threshold of the direct link to the target.

Morphogens can be transmitted between cells in different ways (Fig. 6.7): by active transport through dedicated channels in cell junctions, along the surface of a cellular tissue, or by diffusion in the intercellular matrix. Differences in diffusivities of morphogens play a role in setting the locations of activating and repressing thresholds, but there is no reason for the latter to be less diffusive or otherwise more difficult to transport. The idea of morphogens guiding the expression of genes in different locations was attractive from the outset, and was supported by the Nobelian Francis Crick (1970). He presented an oversimplified picture of a linear morphogen concentration profile due to a source and a sink at opposite ends, and estimated that such a profile could be established in a reasonable time, within a few hours, in a millimeter-sized embryo, but would take a full day in a centimeter-sized animal. This estimate does not change in the more realistic case of an exponentially decreasing profile of a morphogen decaying within a tissue, and is not limited to diffusional transport. Morphogen proteins are not sufficiently stable to travel far, and the essential body organization must already take place in the embryo. The signaling scenario was transformed from hypothetical to proven after the first protein morphogen, *Bicoid*, was identified, forming a gradient along the antero-posterior axis in the *Drosophila* embryo (Nüsslein-Volhard and Frohnhöfer, 1986)[4].

Hans Meinhardt (1982) devised several pattern-forming mechanisms bringing Turing's scheme closer to biological reality. The important points were attention to the polarity of the body plan and the influence of boundaries. Polarity is always present in the body plan of animals with either bilateral or central symmetry, as long as their head and tail ends are different and morphogen gradients have a definite direction. In this way, patterning is tied to the idea of positional information put forward by Wolpert (1969). An additional organizer may be incapacitated by lateral inhibition; thus, implanting such an organizer to incite a hydra to grow another head fails if it is placed close to the existing head. Meinhardt observed that hydra, that belovedly simple animal, has separate organizers at both ends, and their realignment could have naturally produced the dorsal–ventral axis supplementing the antero-posterior axis in bilaterians.

Early work concentrated on one-dimensional patterning along the principal direction of a signal – but far more possibilities exist when there are two organizing centers forming morphogenetic gradients along two different axes. Plain combinations of two feed-forward motifs would divide the plane spanned by two axes (say, antero-posterior and dorsal–ventral) into something like a combination of French and German flags in the left panel of Fig. 6.8. This is, however, far from sufficient to explain the rich variety of locations and shapes of the domains of expression. Notwithstanding the complexity of intracellular interaction schemes, the variety of persistent expression patterns cannot exceed the limit set by intersections of the level sets of the signals. Modifying the form and location of a signal source, e.g., replacing a linear source by a point source, would change only the shape but not

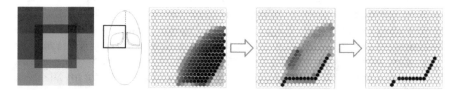

Fig. 6.8 *Left to Right*: "Franco-German flag" with the blackened area affected by the autocrine signal initiated in the central square; *Drosophila* eggshell with the computation domain outlined by the black square; simulated development of the eggshell domain initiating the formation of dorsal appendages

---

[4] This discovery led to Nüsslein-Volhard sharing the 1995 Nobel Prize in Physiology or Medicine.

the topology of the expression domain. Adding more initiating links with different thresholds may increase the number of subdivisions, but domain shapes will always be set by the signal level sets, making it difficult to explain less regular gene expression patterns, such us cusp-like or eyebrow-shaped groups of cells in the eggshell.

Patterning becomes more variegated if external (*paracrine*) signals are complemented by *autocrine* morphogenetic signaling initiated within the embryonic tissue by proteins whose expression is in turn determined by local morphogen levels. Christiane Nüsslein-Volhard (2006), elaborating upon her discovery of the first morphogen, showed how the various autocrine signals determine finer patterning details, like the bands on *Drosophila*, which, even though repetitive, do not in any way follow Turing's recipe for creating periodic patterns, but are created one by one by local signals.

I tried to develop the picture of two-dimensional patterning enhanced by autocrine signaling in a generalized way (Pismen and Simakov, 2011). If there is a single target gene and a single autocrine signal, there are altogether sixteen combinations of their expression in the presence or absence of the autocrine signal. The number of combinations grows exponentially as $2^{2(n+1)}$ with the number n of autocrine morphogens, leading to a great variety of expression domains for the same intrinsic genetic scheme. The diffusional range of the autocrine signal is typically shorter than that of externally supplied morphogens, and if it is expressed in one domain of the "Franco-German flag" it can affect expression of the target in neighboring domains. Thus, expression in the blackened area in the left-hand panel of Fig. 6.8 will be affected by the autocrine signal initiated in the central square. This helps to explain gene expression in domains of convoluted form.

This abstract work was prompted by cooperation with real biologists in a project generously funded by the Human Frontier Science Program. My younger colleague made a brilliant career move from a PhD on the theory of catalytic patterns to experiments with the *Drosophila* fruit fly, which, after serving to fine-tune the laws of heredity, has become the most important "model animal" for development studies. A particular question to be solved was positioning of a narrow and skewed eggshell domain initiating the formation of dorsal appendages. Even experimentalists with a solid theoretical background have trouble with riddles about the inner life of the *Drosophila* fly, and the conjectured signaling and expression scheme kept changing from week to a week. Eventually, we came to a reasonable development scheme including autocrine signaling with the sought-for geometry (Simakov et al, 2012), evolving as sketched in Fig. 6.8. For my part, I couldn't care less about dorsal appendages, so I let my iMac, armed with Wolfram's *Mathematica*, generate random expression schemes with feed-forward motifs leading to a great variety of patterns, with the computer itself selecting those with a non-trivial geometry for my attention. My theory was a flop: the leading paper (Pismen and Simakov, 2011) was never understood nor cited. In our day, researchers are interested in particular details rather than in abstract schemes, and they are surely right. The signaling pathways of development are more variegated than a theorist could ever imagine.

The scaling and robustness problem is an Achilles heel of expression patterns organized by morphogenetic gradients, and it is not helped at all by adding more signaling species. The scale of a signaling pattern is fixed by the diffusion lengths of the morphogens, while in reality development patterns are scaled by the size of an organism; as an extreme example, a mouse and a giraffe have the same number of vertebrae. A variety of mechanisms have been suggested to rectify this contradiction. Some naive attempts suggested doubling a signal by either a counter-propagating signal or a sink at the opposite edge – an arrangement that becomes forbiddingly clumsy in the case of two-dimensional patterning and in the presence of several morphogens. Making decay rates of a morphogen dependent on its concentration levels does not help either; it just deforms the morphogen profile. The scaling problem could be solved by some kind of global control. For example, having morphogen degradation depend on some chemical species present in a fixed amount and uniformly distributed in a developing embryo would automatically make the morphogen gradients scale-invariant, and it could also act in the same way on all morphogens diffusing in different directions. However, global agents require a fast mechanism for sustaining their uniform concentration, and this is unlikely to exist in real tissues.

Naama Barkai and her students have suggested two ways to arrange global control dynamically. One way is to employ a readily diffusible molecule, a kind of "global agent" that "shuttles" the morphogen around, and another protein that breaks the shuttle and releases the active morphogen elsewhere (Ben-Zvi et al, 2008). In another model (Ben-Zvi and Barkai, 2010), the role of a global regulator is played by an "expander" molecule that broadens the morphogen distribution but is repressed by the morphogen, so that, when the morphogen spreads over the entire domain, the production of the expander stops and a size-dependent morphogen distribution is established. Since differentiation of tissues takes place during growth of the tissue, advection and dilution due to division and migration of cells should be important factors. Averbukh et al (2014) presented a theoretical model assuming that cells divide when they feel an increase in the morphogen level, and this leads to the right scaling with size. There are plenty of other papers on this subject, but different mechanisms coming from the same research group are a good indication that Nature might also sustain proportional development of animals, whatever their size, by different means. Genetic patterning must not be directed by morphogens in a strictly hierarchical way, but include dynamic feedback loops.

Besides development processes defining the general body plan and locations of specific organs, there are less prominent ones, generating repetitive regular patterns, such us segmentation, separation of fingers, location of hair follicles, bristles or feather buds, stripes or spots on mammal fur, structuring of insect wings, or the units of compound insect eyes. It appears at first sight that Turing's symmetry-breaking scheme should work straightforwardly here – but Nature rarely concedes to make things simple. Although patterns of fur coloration can be simulated in apparently convincing detail with the help of the universal pattern-forming model, the FN equation (Sect. 3.5), this has nothing to do with their actual mechanism of formation. Segments do not form by symmetry breaking but grow consecutively in the wake of a signaling wave propagating towards the anterior (Pourquie, 2003). Other processes, though based on the same kind of activation combined with lateral inhibition, involve different kinds of communication between cells (see Fig. 6.7). Intricate mechanisms are guessed, disputed, and modeled for the various details of *Drosophila* anatomy, as well as for other model animals, and the search is bound to continue.

As Wolpert (2016) himself admits, *we still do not know the molecular basis of positional information [. . .], nor do we have convincing evidence of how positional values are specified or interpreted.* The hope that particular mechanisms revealed by studies of model animals might be generic is often justified, but it can never be guaranteed that other creatures did not arrive at different patterning mechanisms. The detailed view is frustratingly complex. It is essential when studying human development and physiology, even at the price of sacrificing our mammal relatives to save human lives. It may help to elucidate general principles as well, but it is too often driven by the inertia of academic routine, convenience of experimentation, and availability of funds.

6.4 Mechanics of Tissues

D'Arcy Thompson (1917) was the first to emphasize the roles of physical laws and mechanics in development, rising against the attitude of biologists of his age who were *deeply reluctant to compare the living with the dead, or to explain by geometry or by dynamics the things which have their part in the mystery of life.* It was prophetic at his time to write: *Cell and tissue, shell and bone, leaf and flower, are so many portions of matter, and it is in obedience to the laws of physics that their particles have been moved, moulded and conformed.* Although he is sometimes called the founder of mathematical biology, his methods are far removed from those of modern theories, and he himself admitted his weakness in mathematics.

There is nothing in life that needs more physics than was known in the early 20th century and of which D'Arcy Thompson was aware, though not on a professional level. You don't need quantum mechanics to understand life; the level of quantum-chemical computations lies too deep, and will never be practical for understanding the properties of large molecules, let alone their interactions. But the proper tools for deep study had not yet arrived by then. His was a qualitative approach with clever estimates based on the physics he knew, for example, deducing shapes of

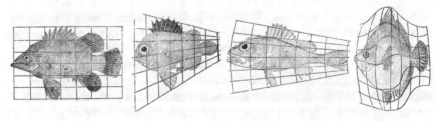

Fig. 6.9 Geometric transformations between shapes of different species of fish (Thompson, 1917)

cells from surface tension – he could not know of myosin motors stressing the cell. The changing shapes of an amoeba presented some problem, so he explained that surface tension is different at different locations; not very persuasive, since it should apparently be negative wherever the shape is concave. His transformations yielding the shapes of various fish (Fig. 6.9) and the spirals of mollusc shells or ram horns were also qualitative, devoid of actual chemical and physical mechanisms. Stephen Wolfram (2002) found it particularly attractive, as it fits well with the digital games of his "new kind of science". Both D'Arcy Thompson and Wolfram are preoccupied with outward features, like visible shapes, pigmentation patterns, or arrangements of leaves, as their approach cannot penetrate deeper into the inner workings of life. Of course, in D'Arcy Thompson's day, there was no other choice.

It appears to be a straightforward task to apply the well-developed apparatus of continuum mechanics to live tissues. The problem lies in complex and largely unknown chemo-mechanical interactions in a jumbled environment of living tissues. Therefore a cell-based discretized approach turns out to be more efficient than direct application of continuous equations (which in any case would be discretized on an arbitrary grid when computing). Deformations and motions of cells are too difficult to observe and model in a three-dimensional setting, but two-dimensional tissues are both common and convenient to handle.

The original blastula (Sect. 6.3) is a two-dimensional shell, and *epithelial* layers of skin, guts, etc., play an important role in a grown organism. Cells densely fill such layers, and their basic arrangement, as in generic two-dimensional patterns, is a hexagonal grid (Sect. 3.4). Of course, the movement of cells, their growth and division, will always distort the ideal regular pattern, but one of its features is robust: the number of three-cell junctions, or *vertices*. On the average, each cell in a layer has six neighbors and six vertices. A four-cell junction is non-generic and highly improbable; even in such an artificial construction as the map of the US there is a single "four-corner" point. A convenient way to study rearrangements of cellular layers is to study the dynamics of vertices. Mechanical laws are introduced in such models in an implicit way by assigning an energy function dependent on deviations from certain optimal values of the area and the perimeter of the cell (Farhadifar et al, 2007). Changing the area will also imply changing the local thickness of the layer if the volume is fixed, or an even more drastic rearrangement of the cell interior if

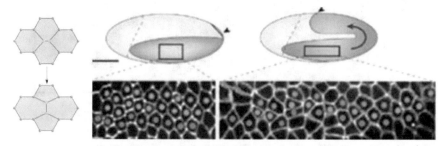

Fig. 6.10 *Left*: Intercalation. *Right*: Elongation of a tissue due to intercalation of cells

the volume changes, and this costs energy. The perimeter increases at a fixed area
if the shape of the cell becomes irregular. Vertices move to decrease the overall
energy, and the derivative of the energy with respect to the position of the vortex is
interpreted as the driving force. Any external force, if present, can be added here,
and it may counteract the tendency of the layer to relax to a regular pattern.

This is, of course, a very rough way to overcome the complexities of restructuring
the cytoskeleton of cells that has to accompany their motion, but it produces realistically distorted cell shapes when combined with cell division and intercalation. The
latter process, shown schematically in the left-hand panel of Fig. 6.10, may naturally
lead to elongation and narrowing of the tissue when a force is applied in a certain direction, or when the cells grow or multiply while laterally confined, as demonstrated
by the transition between the configurations in the central and right-hand panels of
Fig. 6.10. This phenomenon was first documented by Ray Keller (1978). An example of a simulation in Fig. 6.11 (Bratsun et al, 2019) shows how different structures
of a growing carcinoma can be obtained just by varying intercalation probabilities
for cancerous and healthy cells.

A layer of cells can be bent just by contracting one of its surfaces and/or expanding the opposite one. Such a deformation was put forward by Odell et al (1981) as a
straightforward mechanism of *invagination*, the first step in the formation of the inner layer in a developing embryo. Invagination can be initiated at a certain location
by constricting cells on the *apical* (upper, or outer) side and expanding their *basal*

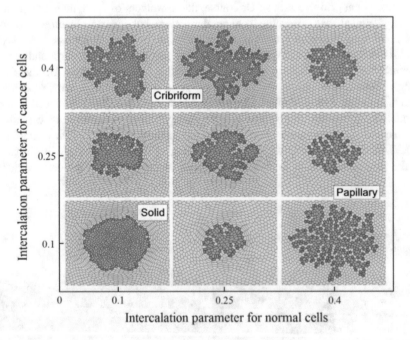

Fig. 6.11 The various structures of a growing carcinoma at different values of the intercalation
probabilities for cancerous (*red*) and normal (*yellow*) cells

(lower, or inner) side (Fig. 6.12, top). Each cell contains a network of contractile microfilaments anchored to the plasma membrane (Sect. 5.4), and stressing them close to the apical surface will produce the desired effect. Odell et al (1981) constructed a mechanical model of this action – but of course, a chemical signal would be needed to trigger it. A sheet of cells can also be folded by a programmed death (*apoptosis*) of cells at a certain location driving its neighbors down (Fig. 6.12, bottom). A two-dimensional view in the right panel (Monier et al, 2015) shows the anisotropy developing near the fold, with the cells elongating parallel to its direction.

The arrangement and motion of cells are correlated with their *polarity*. This term, while having a definite meaning in physics, is loosely interpreted by biologists, sometimes just relating to elongation in a certain direction. There are two kinds of polarization: *vector*, visualized as an arrow pointing in a certain direction, and *nematic*, lacking an arrowhead and therefore invariant under rotation by 180 rather than 360 degrees (Sect. 3.3). Planar nematic polarisation can be associated with anisotropy of tissues, and, indeed, disruption of a uniform planar polarization[5] produces *kugel* (German for "sphere" or "ball") mutants instead of native elongated forms (Gutzeit, 1990). Like everything else, polarity is driven genetically through signaling (de la Loza and Thompson, 2017). In chimeric forms containing both polarizable and mutant cells, this has the character of a phase transition when the fraction of mutant cells exceeds a certain limit (Viktorinova et al, 2011).

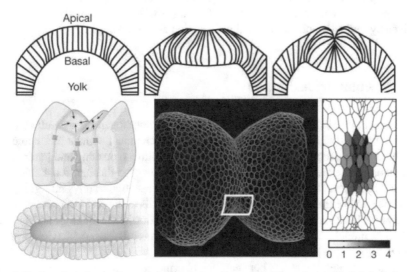

Fig. 6.12 *Top*: Invagination by apical constriction. *Bottom*: Folding of an epithelial shell due to apoptosis: side view *on the left* and a view from above *on the right*. *Shading* indicates the anisotropy of the cells

---

[5] A uniform polarization of a closed shell is topologically impossible, as it must contain defects – but a "meridional" alignment with two aster defects at the opposite ends can be viewed as perfectly ordered for all practical purposes.

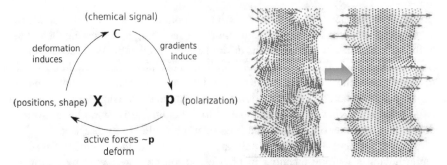

Fig. 6.13 *Left*: Interaction between signaling, polarization, and deformation. *Right*: Evolution of vector polarization from a random distribution (*center*) to a strongly polarized state at the spreading protrusions

Motion is necessarily associated with vector polarization in the given direction, and it also depends on signaling, as sketched in the left-hand panel of Fig. 6.13. The mutual alignment of planar polarity and the direction of motion were observed, in particular, in the development of the *Drosophila* wing (Aigouy et al, 2010) and spread of an epithelial layer into the available space imitating wound healing (Köpf and Pismen, 2013). In both cases, initially misaligned polarizations of cells orient along the flow. As shown in the right-hand panel of Fig. 6.13, polarization is most pronounced near free boundaries of the layer where protuberances are formed, driven by mutually enhancing local polarization and spreading.

6.5 Mechanotransduction

Mechanical regulation complements chemistry in morphogenesis as well as in the functioning of living cells and tissues. Alongside chemical signals that come and go through the plasma membrane and deformations due to pushing and jamming by neighbors, there are specific mechanical signals conveyed through focal adhesions and adherence junctions with adjacent cells. They sense topography and rigidity of a substrate or an extracellular matrix and convey signals to the cell's cytoskeleton and receive its back reaction through actin polymerization, actin linking, and binding adhesion modules, as sketched in the left-hand panel of Fig. 6.14 (Geiger et al, 2009).

A similar interplay exists between neighboring cells connected by an adherens junction. Nascent junctions formed by binding proteins are stabilized by actin cables protruding from the linked cells (Fig. 6.14, top right), while the other component of the junction, formin molecules, initiate and protect actin polymerization (Fig. 6.14, bottom right). These junctions serve at the same time as gateways for intercell chemical signaling (Sect. 6.3).

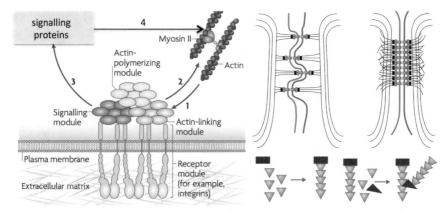

Fig. 6.14 *Left*: Iinteraction between focal adhesions and cytoskeleton. (1) Myosin-dependent contractility affects mechanosensitive proteins in the module linking actin to a focal adhesion. (2) The back effect of the mechanical force felt by the substrate or the intracellular matrix on the actin cytoskeleton. (3) Release of signaling protein molecules. (4) Cytoskeleton-regulating proteins affect actin polymerization and actomyosin contractility, thereby modulating the force-generating machinery. *Top right*: Maturation of an adherens junction driven by tension in the actin fibres, seen as extended lines in the nascent junction on the left. *Red* and *green lines* indicate cell membranes, formin molecules are shown by *black rectangles*, and binding proteins are *blue. Bottom right*: Formins trigger polymerization of actin monomers (shown by *triangles* with the pointed end directed downward). A protein molecule initiating branching is shown by the *black triangle*

Mechanotransduction, the conversion of mechanical forces into structurally relevant information, plays an important role in the various developmental processes, as well as in pathology, such as the spread of cancer (Hoffman et al, 2011). One of the important ways it works, as schematized in Fig. 6.14, is by breaking weak (non-covalent) molecular bonds and hence modifying protein conformations, as shown in Fig. 6.15. The cellular response to forces transmitted through focal adhesions and adherens junctions may be much faster than the response to chemical clues. As mentioned in the preceding section, parameters of vertex models are just surrogates for actual complex interactions.

Ambitious attempts at a theory of *morphomechanics*, stressing the active role of mechanical factors in development, were independently pursued, starting in the

Fig. 6.15 Change in the energy landscape (*left*) and the resulting conformation and function changes (*right*) under an applied force

1980s, by Lev Beloussov and Larry Taber. Beloussov et al (1994) presented the elongation by intercalation caused by an external force as a prime example of an active mechanical response, which is directed towards restoring a naturally stressed state. He went even further with the *hyperrestoration* principle, stating that a living tissue always overshoots it reaction. Taber (2009) presented several examples contradicting the hyperrestoration hypothesis, but both he and Beloussov with their coworkers have examined many mechanical effects in development over the course of more than thirty years of studies.

Beloussov (2015) unconvincingly criticized the positional information concept that establishes relations between cell positions and their fates (Sect. 6.3), and even the very idea of genetically programmed development, while stressing interactions among cells as the principal morphogenetic force. I mentioned before some problematic aspects of the theory of morphogenetic gradients, but its strength is in intercellular signaling complementing externally imposed gradients and driven, like everything in life, by genes. Mechanics could never be as specific and precise, but it plays a subtle role in modifying chemical signaling. Morphomechanics remains outside the mainstream where the shepherds of flocks of students and postdocs are laureled by prizes for untangling chain by chain the webs of protein interactions.

6.6 Growth and Movements of Plants

It is unfair to talk only about animals. Plants are feeding us, directly or through a food chain. But plants are silent, and they don't write books. Plants, like animals, start their life with the merger of an egg and a sperm cell, and meiosis. As an animal egg polarizes to form its head and anus, a plant seed polarizes to form its shoot and root – but here the similarity ends. While the animal's (or its larva's) organs already develop in the embryo and afterwards only grow and mature, plants continue their *phyllotaxis*, the process of generating new *phylla* – leaves, roots, stalks, florets – while they are alive. Throughout their lives, they retain embryonic tissues, *meristems* (Fig. 6.16, left), nucleating *primordia* of these repetitive structures, starting as small undifferentiated bumps on the surface of a plant.

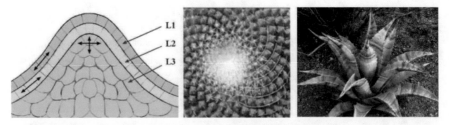

Fig. 6.16 *Left*: Apical meristem of a growing tip with the outer epidermal (L1) and subepidermal (L2) layers and the inner volume L3. *Center*: Dense pattern of two mutually intersecting spirals in the top view of a cactus. *Right*: Spiral pattern of agave leaves

Arrangements of growing leaves and other iterative phylla have long fascinated both natural scientists and mathematicians. The spacing of primordia is determined by the inhibitory action of existing leaves, which is reminiscent of Turing's symmetry breaking mechanism, although it was realized by botanists (Schoute, 1913) long before Turing. An intriguing feature is the pattern of two mutually intersecting spirals, called *parastichy* (Fig. 6.16, center and right), related by Fibonacci numbers. This classical sequence, originating from the counting of breeding rabbits in a book written by Leonardo of Pisa in 1202 (known by his patronimic, *filius Bonacci*), but known in India perhaps as early as the 5th century BC, starts with 1 and 1, while each subsequent member of the sequence is the sum of the two preceding ones: 1, 1, 2, 3, 5, 8, 13, etc. What has this got to do with plants? Douady and Couder (1992) constructed spiral patterns governed by the Fibonacci sequence with the help of a model so simple that it can be explained to a lay reader, and implemented it both by simulations and by experiments bearing no relation to plants.

Botanists had come upon the empirical rule that a new primordium appears with periodicity T near the tip in the largest gap left between the previous primordia and the apex. Couder and Douady assumed that, due to inhibition, new phylla nucleate at a distance R from the apex as far as possible from the preceding ones and are advected from the center due to growth at a velocity V. In this model, all relevant distances should be proportional to R and all times to T, so that these values can be used as length and time units. Accordingly, R/T can be taken as the velocity unit. The resulting pattern should depend only on the dimensionless rate of growth V.

Let the first primordium appear at time 0 to the east of the tip, i.e., at an angle $\phi = 0$. The second should then appear at time 1 on the west side, i.e., $\phi = \pi$ (180°). At time 2, the first outgrowth will be at distance 3 from the center, and the second at distance 2, both retaining their angular positions. When growth is very fast, $V \geq 2$, each new primordium is repelled only by the previous one, so that successive dots move away in opposite directions. In the interval $2 > V \geq 1$, the angular position of a new primordium depends on the two preceding ones, at $1 > V \geq 2/3$, on three, at $2/3 > V \geq 1/2$, on four, etc., at decreasing intervals. As the growth rate decreases, the phylla form a spiral pattern with the number of branches increasing step by step as the growth rate decreases.

Couder and Douady used a dynamic computation minimizing an "energy" dependent on the repulsion strength to compute spiral patterns, but it can be done in a simpler way by determining the locations farthest from already existing phylla at each step. Miraculously, the pattern advances along the Fibonacci sequence with decreasing V in such a way that sharp transitions occur precisely at the above-mentioned values of the growth rate where the number of phylla influencing nucleations changes. For $V \geq 2$, phylla lie on two straight lines going in opposite directions from the center; any two consecutive phylla, labelled by their nucleation time, can be connected by an arc between these two lines, and the labels of consecutive phylla lying on the same line differ by two. Recall first the Fibonacci numbers, 1 and 2. Below $V = 2$, a true pattern of intersecting spirals forms; along one of them, the labels of the consecutive phylla differ by 3, and on the other, by 5 – the next members of the Fibonacci sequence. In a denser spiral pattern with $V < 1$, the

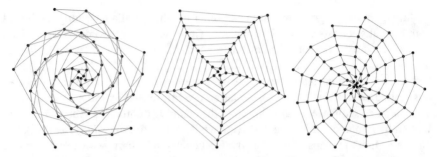

Fig. 6.17 Patterns of mutually intersecting spirals defined by Fibonacci numbers 3, 5 for $V = 1.2$ (*left*) and for $V = 1$ (*center*), and 8, 13 for $V = 0.8$ (*right*). Locations of the phylla are marked by dots. The first of these, at the outer edges of the spirals and deviating from the regular pattern, are omitted, and arcs are replaced by straight lines

respective numbers switch to 8 and 13 (Fig. 6.17), again as they must if this law is to be obeyed, and so it goes, with the spiral pattern becoming denser and computations more tedious as further critical values $V = 2/3, 1/2$, etc., are passed. The number of spirals is conserved within the intervals between transition points, and only the shape of the spirals changes.

This model catches the essence of phyllotaxis, but in reality everything is not as simple, since signaling and mechanical forces certainly play a role in this, as in all other morphogenetic processes. The growth hormone auxin is instrumental in initiating the formation of primordia, and compressive stresses lead to buckling of the plant surface. Newell et al (2008) and Shipman et al (2011) developed a model taking these factors into account and found that Fibonacci patterns are persistent among different growth scenarios, but sometimes come out with imperfections, while transitions between them become less sharp, so that different spiral patterns may coexist within certain parametric intervals.

Auxin, a small molecule so important for plant growth, plays contradictory roles. It promotes growth of roots but suppresses shoot branching, competing with other hormones that take over when the apex of a growing plant, where auxin is produced, is cut, or downward transport of auxin is suppressed (Leyser and Domagalska, 2011), as shown in the three left-hand panels of Fig. 6.18.

In addition to the formation of new phylla and growth by cell division, a plant may grow through cell elongation, which can be directed in specific ways: axially in stems and roots, or forming the flattened structure of a leaf. Expansion of cells may be caused by osmotic flow of water into the cell with a higher solute concentration. This generates what is called *turgor pressure*, regulated by a *vacuole*, as in the right-hand panels of Fig. 6.18. The influx of water extends the cell walls, as in the lower panel. In the opposite case of low solute concentration, water is driven outside, so that the cell wilts, as in the upper panel. Neither can happen in animal cells lacking a cell wall.

Water intake is largely responsible for movements of plants – not traveling movements, of course, but changes in shape that enhance their proliferation and survival.

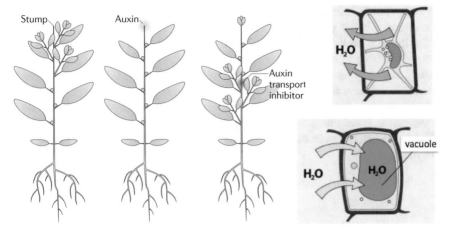

Fig. 6.18 *Left to right*: Shoots branch when the top of the plant is cut; auxin produced at the apex inhibits branching; shoots branch when transport of auxin is suppressed; hypotonic (*bottom*) and hypertonic (*top*) cells with, respectively, high and low turgor pressure and inward and outward water flow

Deformations caused by changes in humidity enhance seed dispersal in wheat awns, pine cones, and other plants. The mechanism is particularly sophisticated in desert plants, which need to scatter their seeds when more moisture is available and they have better chances to germinate, as illustrated in Fig. 6.19 (Harrington et al, 2011). In the dry state, seed compartments, partitioned by *septa*, are covered by *keels* (bottom petals) that serve as protective valves preventing premature dispersal of seeds. When the keel tissue absorbs water and swells, as shown in the bottom panels, each valve, consisting of two halves separated in the dry state but coming into contact when swollen, unfolds in a sophisticated bidirectional way.

The ability of plants to grow permanently by creating new cells of whatever kind they need, their self-sustenance, and the lesser degree of specialization of plant cells

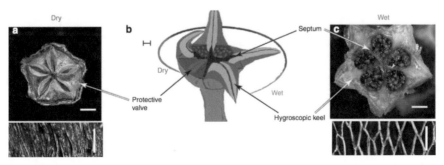

Fig. 6.19 *Left*: The unfolding mechanism of the desert ice plant seed capsule (*top*) and the change of the keel tissue structure from the dry to the wet state (*bottom*). *Scale bars* are 2 mm (**a, c**), 1 mm (**b**), and 0.1 mm in the *lower panels*

as compared to animal cells – all this enhances their ability to survive. Rather than moving to a better place, plants adapt to changes in their environment by getting taller to catch sunlight and extending their roots to imbibe more moisture. They are not as sensitive to radiation (for example, prospering in the Chernobyl exclusion zone), and they never die of cancer, as their rigid cell walls prevent metastasis. Plants do not have a nerve network, and therefore are not in pain when eaten. Lots of advantages – but we still don't envy them. A person in a coma is said to be in a *vegetative* state. Fungi also share many of plants' dubious privileges – but they deserve less attention from our self-centered point of view, as we don't need them for our survival.

Chapter 7
Communication

7.1 Flow Networks

Large organisms need to develop special internal exchange systems to supply nutrients and remove waste. Mere diffusion already becomes ineffective on millimeter distances, and the supply path, proportional to the volume-to-surface ratio, increases with size. This necessitates more effective internal communication highways with a large surface area and transport mechanisms that are more efficient than diffusion. Branched structures, either in the form of what mathematicians call *trees* (because real trees and their roots are like this), or connected into networks, have a very large surface area for a given volume (Sect. 3.6), and therefore work most efficiently when this task is set. Within branches, material is carried by flow rather than diffusion, which is not only faster but can be more efficiently directed to the relevant destinations. Both animal and plant branched structures are graded, becoming thinner as they branch, in accordance with the decreasing flux.

Plants grow flat leaves with a large surface area that capture sunlight. They did not have a nutritional problem while they grew under water, but after venturing onto dry land about half an eon ago they could not survive without procuring moisture and minerals from soil, and had to develop branched supply systems both below and above the ground. Plant roots and stems or trunks, like other phylla, are initiated by apical meristems (Fig. 6.16, left), and elongate and thicken to satisfy the plant's needs, driven by auxin and other hormones. Roots do not form ordered patterns like those discussed in Sect. 6.6, and such patterns are only rarely distinguishable in branches of trees. The distinction between the two kinds of branched structures may blur when roots emerge above ground and trunks penetrate below, which is particularly evident in mangroves growing in waters raised and lowered by tides. The branched root system (Fig. 7.1, left) combines its nutritional function with supporting the plant mechanically, and this is also helped by having a large surface area for stronger contact with the soil. The large surface area of stems serves to support larger numbers of photosynthetic leaves in such an arrangement that they do not shade one another.

© Springer Nature Switzerland AG 2020
L. Pismen, *Morphogenesis Deconstructed*, The Frontiers Collection,
https://doi.org/10.1007/978-3-030-36814-2_7

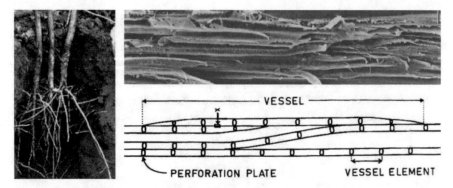

Fig. 7.1 *Left*: Roots of a cotton plant. *Right*: Longitudinal section through xylem capillaries ("vessels") of a palm stem (*top*) and a schematic representation (*bottom*). Vessels are of finite length and their ends overlap. Water moves from one vessel to the next laterally through pits, one of which is marked by a *small cross*

Within this structure, water is sucked up from soil to leaves through numerous capillaries within both branched structures. The way water moves against gravity in tall trees was a riddle until the late 19th century. Henry Horatio Dixon (1914) suggested that it is sucked in by negative pressure arising due to transpiration from the leaves. This idea triggered a controversy when first put forward in 1886 because negative pressure should cause cavitation breaking the continuous path. The liquid cohesion is, however, strengthened in thin capillaries where cavitation is prevented by surface tension (Canny, 1977). Such a continuous set of capillaries is formed in the *xylem* tissue (Fig. 7.1, right) containing both living and dead cells with their walls perforated to produce a continuous pipeline (Tyree and Zimmermann, 2002). A parallel *phloem* system transports sugar synthesized in leaves in the opposite direction to build up the body of the plant.

Animals always depended on external sources of oxygen and food. Sponges solved the problem of material transport in the simplest way by letting water freely circulate through a network of pores in their bodies. Cnidaria also keep the surface to volume ratio high in umbrella-shaped jellyfish or bushy sea anemones. The triploblastic animals which came onto the stage in the Cambrian explosion developed a highly branched vascular blood circulation system in which nutrients and oxygen are transported by *advection* driven by a pulsating heart; this system also serves to integrate body functions through thermal and hormonal regulation and immune defence. In arthropods and molluscs (except cephalopods), it is an open branched structure that empties blood into the body cavity, but cephalopods and vertebrates evolved a closed system that consists of two structures, arterial and veinous, with the heart pumping blood in and out of body tissues through highly branched connecting capillaries (Fig. 7.2). A mirror pulmonary network passes through the lungs or gills to saturate the blood with oxygen. Octopuses and squid are also offbeat in this design, having two separate hearts pumping blood through the gills. The formation of new capillaries in growing tissues or tumors – *angiogenesis* – proceeds by branching of existing capillaries, either by splitting or by puncturing their walls, and concerted spreading of epithelial cells bounding the vessels.

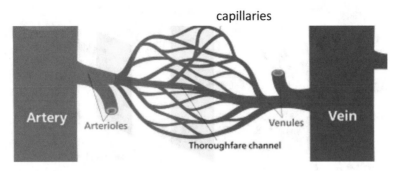

Fig. 7.2 The vertebrate blood circulation system

7.2 Nerve Networks

Animals supplemented material flow and chemical signals carried by blood by faster electrical communication. Diffuse nerve networks have probably already existed in Ediacarans (Northcutt, 2012), but the central processing unit, the brain, is absent in organisms with radial symmetry, and appeared only in bilaterians at the time of the Cambrian explosion. The components of these networks, nerve cells, themselves have dendritic structures attached to their long axons (Fig. 7.3), which allows them to make multiple connections with other nerve cells. Connections in nerve networks are not just nodes appearing in network diagrams (when they are simple enough to be drawn). Those are *synapses*, protein switches which open or close depending not only on incoming signals but on the various chemicals affecting their conformations. A neuron is "fired" in response to synaptic signals; this is a binary on–off operation superficially similar to changing the state of a computer bit, but involves a far more complicated machinery of transmembrane ion transport (see Sect. 5.3 and Fig. 5.7). This seemingly digital operation is not perfect, as it is subject to noise and sensitive to chemicals modifying both synaptic and membrane transmission. Some neurons (in particular, those in visual contours), do not spike but transmit graded electrical signals in the same way that analogue machines operate. This, together with the high connectivity of the network, is responsible for both the advantages and disadvantages of brains compared to computers.

The amount of transmitted information, like material flux, is limited by the physical properties of a transmitting channel, and thicker transmission cables or more efficient media are needed when the information flow is intense, but information networks have to solve more intricate tasks as well. Grading a network is a far more intricate problem than grading the width of channels in a branched flow structure, but this is not the hardest problem of brain design. The speed of information transmission is boosted by encasing axons in a fatty myelin sheath that insulates them and prevents electrical losses. Ion exchange through the axon membrane does not take place continuously in myelinated neurons, as in Fig. 5.7, but only in the gaps between the nodes of the sheath, called *nodes of Ranvier* (Fig. 7.3), so that the prop-

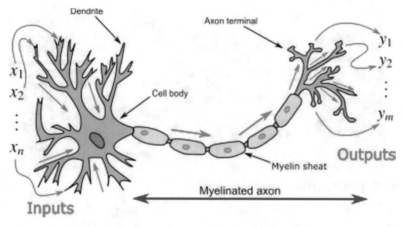

Fig. 7.3 An axon connecting multiple input and output terminals

agating pulse jumps from one node to the next. This innovation appeared in jawed vertebrates (a group containing all vertebrates we know about, except lamprey eels and their relatives, mostly extinct) – but not all neurons are myelinated, not only because of the expense of advanced design but in order not to impede chemical signals where they might be helpful.

Non-myelinated axons, like synapses, are sensitive to chemicals affecting the membrane potential, and even may interact with neighboring nerve cells by *ephaptic* coupling through ion exchange or induced electric field. Most advanced parts of the brain are protected by the selective *blood–brain barrier* separating them from the circulating blood, but nerve cells of old design prevail in the oldest parts of the nervous system, in particular, in the largely autonomous *enteric* net regulating gastrointestinal tracts. This "second brain", as it is sometimes called, sends to the central processor both electric signals through the *vagus* nerve and chemical signals carried by the various messenger molecules affecting synaptic transmission and thereby changing our mood, and sometimes causing us to change mind as well.

A brain is *not* a digital computer but a parallel distributed electrochemical machine that operates in a mixed digital and analogue mode. The analogue component is stronger in the nets responsible for internal regulation, which are in many ways similar to pre-Cambrian nets, but it reaches up to the highest cognitive functions. Antonio Damasio (2018) associates the analogue chemical component with feeling and affect, and digital, with intellect and reason. This is a powerful metaphor substituting for evasive quantitative definitions. Chemistry matters: the computing system of the European Human Brain Project that mimics two hundred thousand neurons and fifty million synapses (still a minute number!) on a silicon wafer would never imitate the "feeling" component. In Damasio's words, *the notion that the human mind can be 'downloaded' into a computer [. . .] reveals a limited notion of what life really is and also betrays a lack of understanding of the conditions under which real humans construct mental experiences.*

Brains can still beat computers in many essential tasks, like image recognition and motor action. The human brain is 10 million times slower than a run-of-the-mill electronic computer, and is therefore at a disadvantage in logical operations, but operates with an amazing speed to solve tasks that are really important for survival. Whereas a well programmed computer beats chess and *Go* champions, a professional player uses circuits inherited from the ancient hunter and honed by training to compute in a split second the trajectory of a ball and actions required to intercept it; a clumsy robot is no match in this skill – so far. An octopus, with its distributed nervous system placing parts of the brain and photosensors in its tentacles, is unbeatable in the speed of computations allowing it to adjust the shape and color of its body to a changing environment (Sect. 3.2).

Computational neural networks imitate the high connectivity of live neural networks and, what is most important, their ability to modify connection strengths in response to activity and experience. Unlike the brain, they are not physical devices but programs, so they retain electronic speed and precision and are free of chemical interferences; their electronic dreams would not be induced by psychedelic drugs. This kind of programming is most successful in machine learning, in pursuit of Artificial Intelligence (AI). Still, its structure is not as flexible as the brain's: it is organized in input, output, and hidden layers, while the brain has an interconnected architecture of which we still do not know much, and which we may never know in detail, as it is certainly unique to each individual, and changes in time. The race is on! AI is developing incomparably faster than our mental abilities (which perhaps even decline as we become tied to computer screens) – but will it ever cross the line between computation and intuition? Is the chemistry of our brain the magic force, the hidden root of what we feel as mind, as soul – or just an obsolete residual of animal development?

7.3 The Bacterial Alternative

Innovations in the design of multicellular organisms were built upon the basic chemistry and structural organization of single cells. Bacteria don't have specialized sense organs but they are able to sense the chemical composition of their surroundings through receptors on their plasma membrane, and are sensitive to temperature and illumination. They cannot do without it, because they need to find the way to a food source or move into a more agreeable neighborhood, just as we do, and they are suitably equipped for locomotion (Sect. 5.6). They are not permanently tied to other cells but they are able to communicate by secreting and sensing certain chemicals, and hence cooperate.

Bacteria form colonies in nutrient solutions, most commonly on surfaces, and can be more tightly integrated in *biofilms*, where they become embedded in a slimy matrix. Interactions between cohabiting bacteria may lead to the formation of clusters similar to the aggregation of particles (Sect. 2.3) or droplets (Sect. 2.5), or pattern formation involving the same kind of activator–inhibitor combinations as

Fig. 7.4 Different shapes of modeled (*left*) and observed (*right*) bacterial colonies

in reaction–diffusion systems (Sect. 3.4). Communicating bacterial colonies can develop in different *morphotypes* (Ben-Jacob et al, 1994), for example, dense or ramified branching patterns like those shown in Fig. 7.4, and they may also spontaneously gain chirality, growing into vortex-like spirals.

The shapes of colonies in Fig. 7.4 resemble branched structures observed in non-living systems (Sect. 3.6); similar pictures would indeed be obtained if bacteria just multiplied by consuming a nutrient diffusing from surrounding space. Yet, a different mechanism involving bacterial interactions is at work here. In the model computations presented in the left panel of Fig. 7.4 (Ben-Jacob et al, 1994), bacteria were represented by walkers moving randomly and multiplying when there is enough nutrient or remaining stationary when they lack food and have no energy to move. Cooperation is imitated in the model by the presence of a barrier to spreading that can only be overcome when more walkers reach it, thereby getting better access to food. The patterns are denser and more compact at higher nutrient levels or for easier spreading imitating a softer substrate.

The most common cooperative mechanism affecting the structure of colonies is *chemotaxis*, which skews the random walk, preferentially directing it along or against the gradient of a nutrient, the temperature, or a signaling chemical. Microbes cannot contemplate and make plans about where to go; moreover, they are too small to sense a chemical gradient. Instead, they sense a change in intensity or concentration over time. Engelmann (1883) proved this by switching a light on and off in a homogeneous solution: bacteria reacted to the change by backing up and changing their path. This implies that they have a kind of a short term memory, which allows them to measure the spatial gradient by sensing a change in time as they move. Some bacteria move in a *run-and-tumble* manner, alternating motion in a certain direction ("runs") with random turns ("tumbles"). If the bacterium senses

that a run is going in the wrong direction it tumbles with a higher probability, and
the other way around.

It is quite comfortable to devise and implement models where "walkers" representing either bacteria or abstract "active particles" move while interacting with
other particles through induced fields or just sensing them in the vicinity. Walkers may also possess geometric characteristics, most commonly, *polarization*. The
bacterium habitually used in experiments, *E. coli*, is oblong; thus, it can be characterized by a *director* aligned with its long axis (Sect. 3.3). Interaction between
polarized particles favoring their mutual alignment is also a feature of many models.

The direction of motion can also be viewed as a geometric characteristic, equiv-
alent to vector polarization. The simplest model of collective motion (Vicsek et al,
1995) contains a single interaction mechanism: the velocity of each particle has to
align with that of its close neighbors. This model has become very popular, being
applied even to such animate assemblies as bird flocks and fish shoals. The advantage (or, depending on your point of view, *dis*advantage) of models of this kind is
that they may produce pictures bearing a superficial resemblance with observations
even when they do not reflect the actual way the system in question operates. Alignment of migrating birds and fish is motivated by hydrodynamics: in this way, they
save propulsion effort. Computing hydrodynamic interactions is a hard task. In the
case of microbes, interactions are carried by induced viscous flow (Sect. 5.6) and
can be computed, though not easily and not precisely, but in the case of fish and
birds, solving the hydrodynamic problem involving inertia and turbulence would be
hopeless. The Vicsek model avoids this trouble.

Models including chemotaxis have plenty of leeway, as there are a great many
ways to assign signaling species, their emission and sensing characteristics, and possibly also the manner in which polarization modifies the motion. But a resemblance
with experimental results, as I already mentioned, never proves a model's validity.

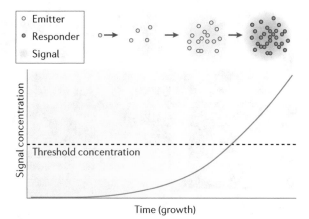

Fig. 7.5 Quorum sensing

Ben-Jacob et al (1994) modified the model discussed above by adding chemotaxis in the field of a signaling chemical, secreted or consumed by bacteria in a way that depended on their level of nutrition. This provided another interaction mechanism that changed the shape of the colony with other parameters kept fixed, but one could find a similar form among the patterns in Fig. 7.4. Designing various models of "active media" became popular among theorists. Some of them produce dazzling patterns, beating anything displayed by abstract expressionists. The review (Marchetti et al, 2013) coauthored by seven prominent physicists from three continents included 280 references and accumulated over a thousand citing papers over a period of six years, of which only about 5% are directly related to biology.

Biologists are more concerned with biofilms where the embedding matrix inhibits motion but chemical interactions are pronounced to a degree approaching multicellular organisms. The integrating mechanism is *quorum sensing* (Shapiro, 1998; Waters and Bassler, 2005). Bacteria detect signaling molecules and change their gene expression, and hence also their behavior, when a certain threshold is passed (Fig. 7.5). Biofilms usually include different microbe species, and bacteria are able to differentiate between signals sent by their kin or non-kin. Cooperation between bacteria, either genomically related or not, involves the metabolic costs of signaling, which are rewarded by the benefits of division of labor that leads to more efficient proliferation, and of collective defence against invaders or antibiotics. Interaction between bacterial species may even go as far as exchange of DNA, the bacterial analogue of sex.

Division of labor may also develop among bacteria of the same strain. In the experiment by Liu et al (2015), bacteria near the outer fringe of the biofilm had better access to externally supplied nutrients, while those in the interior were starved. The peripheral bacteria are, however, more vulnerable to outside threat. When attacked, they die to protect their interior kin, who can then get access to food and proliferate, replacing peripheral cells, as illustrated in the left-hand panels of Fig. 7.6. The divi-

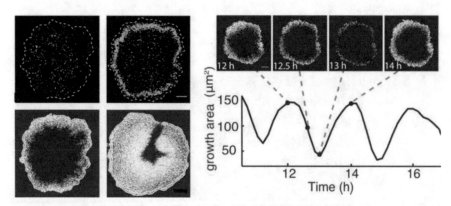

Fig. 7.6 *Left*: The response of a biofilm to an attack. Green dots in the upper row indicate the death of cells and white dots in the lower row, proliferation before (*left column*) and after (*right column*) the attack. *Right*: Oscillations of the growth area (shown in white in the respective pictures)

sion of labor between well-fed defenders and undernourished folk inside increases resilience to outside threats – but the bacterial community also contrives to prevent mass starvation in quiet times. Outside cells periodically stop proliferating, letting more nutrients penetrate inward, and this leads to growth oscillations of the kind shown in the right-hand panels of Fig. 7.6.

Cooperation among cells is indeed the bacterial alternative to integrating into a multicellular organism, and might have been the precursor of the transition to multicellularity. But is it as efficient? Bacterial chemical communication is not directed to specific cells. It lacks a *structure*. Different bacteria care for themselves rather than for the biofilm as a whole, which may benefit from their actions only unwittingly, not unlike human society benefitting from the egoistic behavior of its members. Bacterial, like human, societies may suffer from the "tragedy of the commons", overconsumption of public goods. And bacteria would not commit suicide, emulating the apoptosis of organismal cells.

Bacteria may specialize within a cluster or a colony, and, going further, specialization and cooperation among clusters may also be possible. Biofilms possess some mechanical integrity but it is hard to imagine them developing a kind of muscle without integrating into a single organism. Muscles should be coordinated and controlled, and control implies not just the ability to communicate, but communications sent to a precise address. This kind of control is directed by genes in cells, by nerve networks in animals, and by social or government regulations in animal bands or human tribes, communities, or polities, but it cannot be maintained in an unstructured assembly.

7.4 Collective Advantage

Collective interactions without integrating into a single body develop not only among cells but among multicellular organisms. I mentioned the hydrodynamic advantages of fish shoals and bird flocks. Those are temporary and weakly interacting assemblies, generally defined as *swarms*. It is also often said that swarming provides defence from predators, so-called "safety in numbers". I doubted it when watching a film by David Attenborough showing predator fish and dolphins feasting on a shoal of herring, until "there were none". Land predators are commonly not as numerous and have a limited appetite, so that brawny wildebeests or deer may indeed feel safer in a herd because it is their weaker relatives who will be eaten. More likely, kin vertebrates herd and swarm just because they have common tastes.

Stronger cooperation is rare among animals, and in mammals, not counting humans, it goes no further than extended family groups hunting or foraging together. Among insects, only ants, termites, bees, and some wasps have developed tightly structured communities. It is very hard to overcome the egoistic pull of individual selection. A common theory says that altruism may be driven by kin selection propagating genes coding self-sacrificing behavior. The prominent biochemist John Holden reportedly said that he was prepared to lay down his life for eight cousins or two brothers. He wouldn't, of course. The evidence shows that it doesn't work in practice.

Group selection, with more tightly cooperating groups growing and displacing their rivals, may have contributed to the evolution of sociality in early hominids, but evolution of social insects rather followed individual-level selection, *from queen to queen, with the worker caste being an extension of the queen phenotype* (Wilson, 2012). Ant and bee queens stepped on an unusual pathway of evolution. Whereas other animals gradually improved the body design of coming generations, the queens honed instead the self-organized interaction network within their extended bodies distributed among their progeny. This overcame the limitations of insect anatomy, which does not allow them to grow to a vertebrate's size.

It is quite common to marvel at the clever organization of ant, termite, and bee communities. They are, of course, as much unlike ours as insects are unlike mammals, but these creatures with diminutive brains have managed to build clean and ventilated nests, collectively forage, care for their larvae, even domesticate other insects, and survive in this way for a hundred million years. Ants, as well as other social insects, are indistinguishable and mutually replaceable within their caste. There is no one in charge in the anthill; the queen does not rule, she just bears her progeny, replacing short-lived worker ants. Not unlike bacteria, ants self-organize by interacting chemically, by emitting or sensing pheromones, or visually and mechanically, when encountering other ants. These interactions are not specifically addressed, ants distinguish their kin and nestmates from a stranger but don't say "hello, John, nice to see you again": any ant in a particular colony is like any other ant. What counts, is the frequency of encounters that provides "quorum sensing". There is no persistent specialization. Any worker ant can carry out any work, although the response to stimuli varies with age, and it may be possible to discern an elite of "exceptionally active or entrepreneurial individuals" (Oster and Wilson, 1978).

Workers are mobilized for urgent tasks, be it food delivery or nest repair, and in case of trouble either hide in the nest or attack the invader, but no one tells them what to do, and they have to decide it independently. This calls for a sophisticated communication language. Decisions on work allocation are taken in response to a *pattern* of inputs, either attractive or repulsive, including environmental stimuli, frequency of encounters with workers busy on particular tasks, and visual and olfactory signals. Relevant information may assume specific forms; thus, ants mark trails to food sources using pheromones, and bees indicate them by "dancing" movements. Processing this involves a kind of mental computation, and requires memory. The cognitive power of an insect's brain makes feasible what a microbe swarm could never achieve.

Still, all actions are "robotic"; it would be quite feasible to program automata to operate in this way. This would not need a billion euro budget of the European Human Brain Project (Sect. 7.2). It can also be imitated by a not too complicated mathematical model (Gordon, 1999). Moreover, cooperation is not perfectly coordinated: among a group of ants carrying a load, one can observe individuals pulling in opposite directions.

How far can swarms and animal communities develop? Stanislaw Lem (1961) imagined a mighty sentient ocean, but neither he nor his ill-fated astronauts, who were sent to investigate it but were instead investigated themselves, had any clue of its nature. "Swarm intelligence" in the literal sense is impossible. Diffuse communication without precise addressing, either chemical or of any kind Lem's ocean might possess, cannot imitate an interconnected nerve network. Going a step higher, can social insects develop a collective conscience similar to human's – perhaps on some planet in the galactic depths? Not until members of the colony become distinguishable, as humans are, and individuality requires more internal complexity, sustained by a larger brain than can be afforded by an ant. The insect community is as poor an imitation of an intelligent (not necessarily sentient) mammal as a biofilm is of a multicellular organism.

Humans do possess "swarm intelligence" in a wider sense. Human society knows and can accomplish much more than any individual – but only because any human can be addressed individually, and only while a community is properly organized: when it is literally a swarm, like Facebook "friends", nothing but chat and fake news comes out. Even totalitarian societies striving for robotic unity in pursuit of their great common aims, whatever they are, must distinguish individual people to ensure that everyone is doing the assigned job and does not deviate from the straight and narrow path. How would dictators of the mid-20th century enjoy our connectivity! They couldn't even dream of ideas of connecting people by internet through chips inserted in their brains (Chorost, 2011). Ants and bees do not need secret police, there are no dissidents among them, and neither would there be among us if this dream came about.

7.5 Studying Life

Bodily communication networks, either electrical, chemical, or mixed, not to mention social interactions that cause us to feel delight or pain, all this on top of the complex communication webs within cells and among cells aggregated in tissues, appear at first sight to justify a holistic attitude, going also by the modernized name of "systemic thinking". However, real scientists do not think in this way, and not because they are narrow-minded. They see the abyss of underlying complexity, but this complexity is waiting to be unraveled rather than admired. This is what is being done day by day, finding out what this or that molecule is doing, and how it is affected by this or that gene, and how this or that link in the metabolic or signaling network can be influenced, for good or bad. It is boring to read papers coauthored by blown-up teams, still less to extract the relevant details from them, but sometimes they lead to Nobel Prizes for their leaders, and sometimes they lead to novel cures, and then news, of course, stripped of boring essentials, reach the popular press. This is the research that gathers the bulk of science funding, and it is thanks to this invisible work that we live on the average about twenty years longer than hundred years ago, even though some directions may not seem to be practically relevant at all.

But what about deep general insights? We have seen them in the past; even the nature of heredity was unknown hundred years ago, not to mention the nature of signaling and nerve transmission. However, no Newton of biology will ever come with precise quantitative laws. Perhaps the only general theoretical principle applicable to all communication networks is the necessity of combining activating and inhibiting signals. We can only guess whether Turing (1952) realized the general significance of this principle in his celebrated paper, which treated a very specific problem and ended on a sad admission that biological phenomena are very complicated (Sect. 3.4). Well, we know they are. He himself helped to activate his country's eventual victory by inhibiting German communications, and, in return for this service, was dismally inhibited by the homophobic establishment of the time.

Notwithstanding all the complexity being unraveled piece by piece by biologists, theorists cannot avoid the temptation to model life, and not just on the level of the *Game of Life* (Sect. 2.5). In Sect. 5.4, we talked about the dynamic tangle of the cytoskeleton and in Sect. 5.6 about the clumsy way things crawl. Can this be modeled by applying well established mechanical laws? Yes, certainly, and it has been attempted not once, but the problem is that it is possible to imitate superficial features of observed phenomena by constructing a model that has nothing to do with reality. Attempted models of the mechanics of cytoskeletal rearrangements and cell motion range from the most detailed, imitating the motion of every actin monomer and every myosin motor in the network on a voracious computer, to the simplest coarse-grained models. The former approach cannot be realistically extended to an entire cell, and even on this level, molecular detail, including intricate chemical signaling, cannot be accounted for.

The most sophisticated coarse-grained model is *poroelastic*, originating in the theory of fluid-saturated porous soils (Biot, 1941), which applies separate mechanical equations to the elastic filament network and viscous cytosol. This model leaves aside the permanent restructuring of the live actin–myosin network, as well as the crowded cytosol environment, but it is still computationally difficult, and contains unreliably measured parameters. Simplification goes further, and with a brighter illusion of success. The magic word here is "activity". By treating the interior of a cell as an "active medium" with whatever properties might be helpful, whether viscous or elastic, and polarized in a suitable way, any kind of motion can be imitated. It is rather easy to use one of the modifications of generic models to produce a moving active spot and interpret it as a moving cell, and, whenever needed, add details to better imitate reality. However, a visible correspondence with experiment does not mean that the mechanism built into the model is indeed the one operating in reality. Recall the model of multiplying droplets (Zwicker et al, 2017) in Sect. 4.4. It is simple and elegant but I doubt that the authors seriously thought that the first self-replicating cells might have been formed without being protected by a surfactant sheath.

Mechanical theories become more reliable on a macroscopic level, considering individual cells as elementary units, either incorporated in a tissue or crawling on a substrate, but here again, the difficulty is in incorporating chemical signaling. Modeling is more reliable in *biomorphic* processes involving soft materials but lacking the intricate interactions of living cells. We come to this in the next, concluding chapter.

Chapter 8
Biomorphic Technologies

8.1 Uniquely Human

What makes us unique? Consciousness is an ill-defined entity, as vague as mind or soul. Intelligence is something more concrete; although it also has various definitions, it can ostensibly even be measured. Hence, the Intelligence Quotient, or IQ. Dolphins, whales, elephants, cephalopods, even ravens and parrots are reputed to be highly intelligent – try to administer them an IQ test (of the kind you pass with flying colors if you had learned determinants)! David Stenhouse (1974) defined intelligence as *adaptively variable behavior within the lifetime of the individual*. This greatly widens the circle of intelligent creatures: certainly, all vertebrates are included, and not only them, for they move, and they learn to get food and avoid trouble. Anthony Trewavas (2014) argues that plants are intelligent as well, notwithstanding their *vegetative* state and sessile way of life, after modifying the above definition to *adaptively variable growth and development during the lifetime of the individual*. Plants do indeed exhibit adaptive plasticity but the notions of learning, memory, and individuality have to be shrewdly extended to be applicable to plants. Still more leeway with consciousness: it is up to the personal taste of a neurologist or a philosopher how far to extend it down from humanity.

Intelligent as they might be without brains, plants are able only to grow and reproduce. Whales have larger brains than us (though this does not necessarily make them cleverer), but they made a retrograde evolutionary choice when they left dry land and got rid of limbs to emulate fish: technology cannot develop under water and without appendages fit for manipulating things. They, as well as other sensible mammals, cannot be denied consciousness. Octopuses are dexterous, perhaps one day after we are extinct they will venture out onto the *terra firma* where they will discover fire and start making tools, or creatures like them on another planet will do it. The elephant's trunk is a useful implement though not handy enough (pun intended), but elephants are satisfied with a vegetarian diet and deter predators by their bulk, so they don't need any gadgets. Ants are too small to handle fire without burning themselves. Monkeys' grasping hands with long flexible fingers plus vul-

Fig. 8.1 *Left*: Leonardo's drawing of a flying machine. *Center*: A reproduction of the Golem in Prague. *Right*: An automaton writing a letter in the Swiss *Musée d'automates et de boîtes à musique*

nerability plus curiosity plus befitting ambience is what has prompted our unique achievement – *technology* that has allowed us to rule (or misrule) the planet.

Technology set us aside from Nature. The wheel came as an entirely novel way of locomotion, and remote action by spears and arrows (later bullets) as an innovative method of predation. Nevertheless, inventors understood that Nature is more resourceful and ingenious than they are, and tried to imitate it. An ancient aspiration was to imitate bird flight. Not one unlucky dreamer fell to earth, and not, like Icarus, because sun melted wax fastening their wings but just by their heavy load. The great Leonardo was shrewd enough to draw bird-like flying machines (Fig. 8.1, left) rather than building and trying them. The success came only when unnatural ways of flying were invented: first, a hot air Montgolfier balloon, then a fixed-wing propeller airplane. Another level of technology was needed to build *ornithopters*, and even then mostly bird-size drones, more for entertainment than for practical needs.

Mechanical automata, often imitating human shape, have been imagined, or constructed, or faked throughout human history, from Greece to China, attributed to gods or sages or built by skilled artisans, sometimes magic, like Daedalus' statues, the Hebrew Golem (Fig. 8.1, center) or Frankenstein's monster, sometimes really workable, driven hydraulically, pneumatically, or by clock-like machinery. Like the birdlike machines, these ingenious automata were superficial semblances of natural forms (Fig. 8.1, right), in no way related to their inner working, of which very little was known at the time. They remained toys for the rich, never put to practical use, but Aristotle professed in his *Politics* that automata would eliminate the need for slaves, and they eventually did, enslaving us in a more benign way, as we cannot live without them any more. The modern appellation "robot" was coined by Karel Čapek (1920) from the Slavic root for *work*.

Our separation from Nature peaked following the Industrial Revolution. Metal, so much unlike soft natural tissues, became the major construction material. Ubiquitous automats ceased to imitate a human form. Fossilized sunlight mined underground, rather than wind and water power and more recent sunlight accumulated in wood became the principal source of energy. Electricity, used by animals for internal communication only, became the universal tool for power transmission. Nuclear energy emerged as the power source ultimately detached from and hostile to the chemistry of life.

This started to change late in the last century, accelerating in our own time. It would be impossible to imitate Nature in metal or concrete, but softer materials on a human scale are more suitable for this purpose. Although "tech" generally refers in the popular press to all kinds of computer programming, biomorphic technology has nothing to do with simulated "artificial life", computer games starting with cellular automata (Sect. 2.5) and evolving to an increasing sophistication that prompts postmodern solipsists to doubt whether we ourselves are material creatures or hackers' play. The real progress is in working with materials that can be felt and molded, and in the ways to morph them so as to imitate those employed by Nature.

8.2 Malleable Materials

Biological materials are distinguished by their complex composition and partial ordering that can be witnessed on a *mesoscopic* scale (Sect. 3.2). They are often soft to touch, as their integrity is sustained by weak intermolecular forces. The first material of this kind that already entered technology in the 19th century was vulcanized caoutchouc – natural rubber, followed in the 20th century, first, by synthetic rubber, and then, by a variety of materials synthesized via polymerization of small organic molecules. The portmanteau term *soft matter* was coined by Pierre-Gilles de Gennes who was awarded the 1991 Nobel Prize in Physics *for discovering that methods developed for studying order phenomena in simple systems can be generalized to more complex forms of matter, in particular to liquid crystals and polymers.* Besides polymers and liquid crystals, this term has been applied to *colloidal* materials characterized by a mesoscopic structure or physical properties intermediate between those of solids and liquids.

Polymers became ubiquitous in the 20th century. Plastic, first celebrated for being affordable, ductile, and resilient, became for the same reasons a scourge of garbage, spreading to the open oceans. There is nothing biomorphic about these materials, even though they are based on carbon: to deserve this appellation, the material should at least be changeable after it has been manufactured, and it should be capable of responding in some way to signals and the environment. This minimal requirement is answered by *shape-memory* polymers. We have already met one in Sect. 3.3 – liquid crystal elastomers. They are not alone in their ability to deform when actuated, imitating muscular action. Isotropic polymers changing volume upon any kind of a phase transition qualify as well. You may already be sleeping on a "shape-memory" pillow filled by a foam that adapts to your sleeping position – but this is a misnomer. Any material deforms under mechanical load; foam deforms more, and "memory foam" is a viscoelastic material that rebounds very slowly, so it retains a deformed shape for a while. A proper shape-memory material should have at least

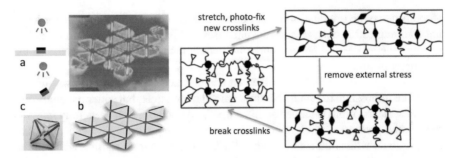

Fig. 8.2 *Left*: (**a**) Folding along a heated stripe. (**b**) The pattern prepared for folding into an octahedron (*bottom*) and its heating pattern (red when hot, green when cold). (**c**) The resulting folded shape. *Right*: Schematic view of reshaping by the photo-induced formation and breakage of crosslinks

two *equilibrium* shapes and be able to switch reversibly from one to the other upon an actuation that triggers a phase transition.

Polymers change their volume when they crystallize or undergo a glass transition. In both cases, the control variable is temperature: a polymer is a malleable elastomer at higher temperatures (but still below its melting point) and becomes glassy or crystallizes as it is cooled down. There are also materials capable of undergoing several phase transitions and hence "memorizing" several shapes (Xie, 2010). Fast uniform heating is not easy to achieve. It can be done, for example, by incorporating magnetic particles in a polymer and heating them inductively – but reshaping under non-uniform heating is more interesting.

When only one side of a strip is heated, it bends as in the lower right panel of Fig. 3.9. More varied deformations can be triggered by elaborating a heating pattern. In this way, Lee et al (2015) ingeniously designed the origami folding of a flat sheet. They printed black stripes on a transparent sheet, which were heated above the glass transition point under illumination, while the transparent pieces remained cool. The black stripes, shrunk and becoming flexible, bent in the only way they could, so that the adjacent transparent pieces folded at a certain angle. Folding into an octahedron, as shown in the left-hand panels of Fig. 8.2, was not a straightforward task, and had to be done by stages. Self-folding is not just a game, it is useful for encapsulation and delivery of drugs, for photovoltaic power applications, and much more.

A more radical way to attain shape memory is to change the internal structure of a polymer with the help of molecular switches providing additional temporary crosslinks, as sketched in the right-hand panel of Fig. 8.2 (Lendlein et al, 2005). A polymer is stretched and the formation of new crosslinks is induced by ultraviolet (UV) illumination. After the external stress is relaxed, the polymer shrinks somewhat but is prevented from returning to the original shape by these crosslinks, and only retains it after they are cleaved by UV rays of a shorter wavelength. A polymer sheet should be illuminated from both sides to keep the stretching and shrinking uniform. This may look like a rather non-biomorphic operation, but we should recall that it is by modifying the internal structure, in particular, by creating and breaking crosslinks in the cytoskeleton, that cells deform and move (Sect. 5.6).

More natural ways to change internal structure than UV illumination can be implemented in *hydrogels*, networks of hydrophilic polymer chains imbibed with aqueous solutions. This material mimics the humidity-induced swelling and shrinking of plant cells (Sect. 6.6), and is also responsive to other environmental factors, like temperature and ionic strength. These capabilities have found a variety of applications (Ionov, 2013). One of them is the smart release of drugs. A microscopic hydrogel particle imbibed with a water-soluble drug is swollen at normal body temperature but shrinks at the higher temperature of an inflammation site and releases the drug. Deformation of hydrogels can also be graded at different locations, causing the material to bend. A variety of shapes can be created in layered or anisotropic hydrogels. Plants also make use of this feature, as we have seen in Sect. 6.6. Once again, none of this yet amounts to shape memory, but different temporary shapes can be created in the same way as sketched in the right-hand panel of Fig. 8.2, but creating temporary crosslinks using milder chemical interactions involving crystallizable or complex-forming side-chains and actuated by temperature or chemical reactions (Lowenberg et al, 2017).

8.3 Digital Morphogenesis

In our age, when computers control all kinds of manufacturing, it is quite natural to ask for their help in shaping malleable materials. The ancient fabrication method by removal of excess material reigned supreme from *Homo habilis* sharpening a stone flake, to a sculptor envisaging the future statue in a piece of marble, to an advanced photolithographic machine carving integrated circuits from silicon wafers. A new way of *additive manufacturing* emerged in the late 20th century and matured as 3D printing in the early 21st. Since deposition is invariably computer controlled, it might just as well be called *digital morphogenesis*.

If you search for this term on the Web of Science, two unrelated themes come into view. First, is the formation of fingers in animal development. This reminds us

Fig. 8.3 *Left*: Anthozoa dress. *Right*: Regenerating fabric system. *Yellow arrowheads* point to newly-made and assembled silk

that the origin of the digital world is in counting on fingers, which also brought us the decimal system, before computers turned up, with their ability to handle digits ultrafast. Another theme, the one that takes the entire top page when googling, refers to computer-aided design and directs us to some marvelous creations.

The 3D-printed *anthozoa dress* by Iris van Herpen and Neri Oxman (Fig. 8.3, left) imitates marine polyps in its texture. A further bold step merges natural and digital morphogenesis in a wearable microfluidic *Mushtari* structure, inhabited by two bacterial communities: photosynthetic microbes convert sunlight into nutrients for the heterotrophs, which can in turn produce desirable compounds, like scents or pigments (Oxman, 2017). Hardly anybody would wear such a dress, but (deviating from the main theme of this section) bacteria have been mobilized for the more mundane task of repairing fabric. Ido Bachelet and his coworkers (Raab et al, 2017) made a first attempt at hybridizing the fabric with a biofilm. In this symbiotic arrangement, the fabric provides a supporting scaffold for the biofilm, while bacteria return the favor by synthesizing a silk protein to repair the fabric when they feel its wear and tear. In a similar perspective, this may be extended to self-cleaning and protective functions.

Artists are good at inventing catchy titles – but the appropriation of the term "digital morphogenesis" is somewhat unfair. Using computers to create forms is now ubiquitous, and it is not so new. Without computer-aided design, it would be too difficult if not plain impossible to design the fantastically curved and arched structures of Frank Gehry and other "deconstructivist" architects, so unlike the boring rectangular boxes of mid-20th century "modern" architecture. It has also made it possible to design the curved aerodynamic forms of our cars, notwithstanding the general conservatism of the automotive industry. And so only a few isolated artisans remain to create forms with their fingers alone, unaided digitally.

Computer-aided design has also led to additive manufacturing through 3D printing guided by programming, which is developing by applying sophisticated microfluidic techniques to integrate different materials into emerging forms (Keating et al, 2016). The range of applications is amazing. Tal Dvir and coworkers (Noor et al, 2019) 3D-printed a rabbit's heart using fat cells reprogrammed to become pluripotent stem cells and differentiated to cardiac tissues. One prospect here, when hopefully extended to humans, will be to eliminate queues for heart transplants, and, in general, make it possible to regenerate any organ without the risk of immune rejection. At another extreme, the Dutch company Concr3De has offered to reconstruct damaged sculptures from the burnt Notre Dame cathedral using 3D printing.

Smart gadgets made of form-changing or shape-memory materials often have a layered structure, and are most conveniently manufactured by 3D printing. Since they are meant to acquire dynamically controllable shapes when actuated, the term has mutated to 4D printing, incorporating the time dimension. This catchy term should appeal to physicists who dream of multidimensional worlds and Einstein's relativity but are stuck in Earth-bound research. Of course, only the spatial form is printed, but the programmed changes in time become inbuilt properties of the biomorphic material.

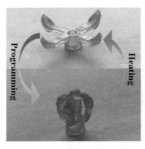
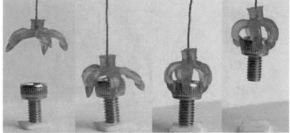

Fig. 8.4 *Left*: The printed (*top*) and actuated (*bottom*) forms of a gripper. *Right*: Manipulating a screw

The way layered plates bend can be predicted straightforwardly, and goes back from modern smart materials to the bi-metal strips of Stepan Timoshenko (1925). If metal A expands more than metal B when the temperature rises, the strip will bend with A on the convex side and B on the concave side. Deformation of shape-memory polymers is, in principle, similar – but leads to more precise and sophisticated forms, because these materials can be tuned by selectively modulating their internal structure. Martin Dunn's group (Ge et al, 2016) carried this out with the help of computer-controlled UV illumination of the polymer as it was 3D-printed, which enabled precise modification of its local structure and, hence, mechanical and transformational properties. Figure 8.4 shows a robotic gripper 3D printed in this way. The technique enables one to fine-tune different stiffnesses in the joints and tips of the grippers for better contact with a manipulated object.

Gladman et al (2016) worked with softer hydrogels imitating plant tissues. They printed patterns using hydrogel ink with imbedded fibrils, which aligned when passing the deposition nozzle, as shown in the left-hand panel of Fig. 8.5. This makes the material anisotropic, so that it swells along the filament length rather than uniformly in all directions. The direction of bending is determined, as in the classical Timoshenko strip, by the way the swelling layer is placed. In the central panel of Fig. 8.5 it is placed on the top or bottom sides of the two sets of filaments, oriented perpendicularly to one another. Upon actuation, this leads to a saddle-like form. In this

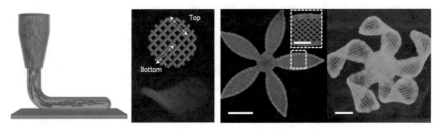

Fig. 8.5 *Left*: Alignment of fibrils in the printing nozzle. *Center*: Grid pattern deforming into a saddle shape. *Right*: A flower-like shape printed and actuated (*scale bar* 10 mm). The *inset* shows the grid orientation in the printed petals

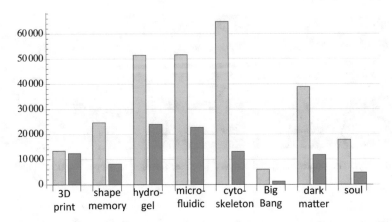

Fig. 8.6 The number of publications on the chosen topics in Web of Science (total in *yellow*, last five years in *blue*)

way, a variety of shapes can be created, like the flower-like forms in the right-hand panel.

These are just two fine examples out of a multitude. I opened Web of Science on Sunday, 4 August 2019. It is a great tool for finding a needle in a haystack or a pearl you know where. At least, the bulk of the trifling stuff offered by Google does not make it into the data base with the access paid by universities, restricted to peer-reviewed papers in journals of some repute. This time, I entered it just to check bulk publication counts for topics mentioned in this section and, for comparison, some others, shown in Fig. 8.6. The number of papers containing "3D print*" (with the asterix standing for any ending) is relatively modest, but most of them were published in the last five years.

8.4 Nanotechnology

Going down the linear scale, biomorphic technology approaches structures commensurate with the light wavelength. The prefix *nano*, a vertex of the trendy nano-bio-info triangle, is, as a rule, generously extended from the nanometer to whatever is less than a micron, and diffraction of light is an important capability of materials made of particles in this range. The number of publications containing this prefix, 1 605 946 in total, 702 958 in the last five years, would dwarf the above bar chart, although of course, only a minor part of them are relevant to nanotechnology.

I have already mentioned in Sect. 3.2 chameleon-like changes of color, which are reproduced by colloidal crystals, also used as optical sensors. Manufacturing these devices requires precision, since they should comprise identical particles arranged in a well-ordered way. This comes naturally in atomic and molecular crystals, as atoms and small molecules of the same kind are indistinguishable, but manufac-

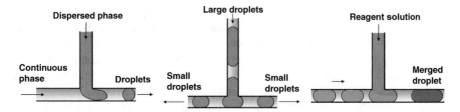

Fig. 8.7 Ways of producing monodisperse droplets in a microfluidic T-junction

turing monodisperse particles of nano- or micro-size in large quantities and at a reasonable price and arranging them in a perfect order is a minor feat of 21st century technology. I recall one of its creators saying in a workshop talk: *I don't know how good we are.* A crystalline texture is viewed under a microscope, either optical or electronic, depending on its scale, but always with a limited field of view. Flaws are so rare that they are never caught within this window; if there were any, one would need to scan the entire macroscopic structure to spot them.

A common method of mass-producing monodisperse particles is to form droplets in a microfluidic T-junction (Fig. 8.7); they come out identical, and can be further solidified. Such nano- or micro-size droplets can also be made with a varied composition and used for many other purposes, such as drug delivery. Of course, manufacturing microfluidic appliances is also a precision task, which is now mastered for more refined uses, going as far as carving sophisticated networks of miniature chemical reactors into a single integrated circuit; there is a special prestigious journal suitably called *Lab on Chip*. Microfluidic devices are used, in particular, to manipulate single cells.

Self-assembly of particles into colloidal crystals of a desired structure is facilitated by attaching molecules that can form specific bonds. DNA with its comple-

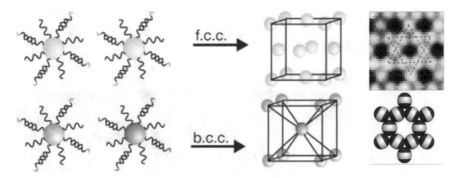

Fig. 8.8 *Left*: Self-assembly of face-centered cubic (*top*) and bulk-centered cubic (*bottom*) colloidal crystals by nanoparticles linked, respectively, by identical and different DNA strands distinguished by the colors of the linking groups (green links to green in the upper, and red to blue, in the lower picture). *Right*: A kagome lattice assembled from Janus spheres

mentary mutually linking bases serves well for this purpose[1]. Attaching suitable DNA strands forces nanoparticles to assemble into colloidal crystals of a specific structure, as shown in the left-hand panel of Fig. 8.8. Another way to stimulate crystallization is to use *Janus* particles which, like the eponymous Roman god, have surface patches with different properties. The spheres shown in the right-hand panel of Fig. 8.8 are hydrophobic near the poles and hydrophilic around the equator, and, to reduce energy, they assemble in the aqueous solution in such a way as to hide the hydrophobic areas, forming a *kagome* lattice (Chen et al, 2011). Still more varied crystalline forms can be built up from "colloidal molecules" pre-assembled from nano-size spheres (Morozov and Leshansky, 2019).

Individual nano-size particles are used for smart drug delivery. I mentioned in Sect. 8.2 how a hydrogel particle would release a drug when it feels the heat of an inflammation, but targeting can be made still more precise. Cancer cells are the most relevant target; we know that killing them by chemo- or radiotherapy or removing them surgically brings nearly as much suffering as the illness itself. Ideally, the drug should be released in a controlled manner in the tumour, minimizing harm to healthy tissues. The most promising way is to carry a drug within a *micelle* protected by a lipid layer similar to that of the cell plasma membrane. Targeting can be made precise by attaching surface ligands to the micelle that would attach to receptors preferentially expressed on the surface of cancer cells.

Furthermore, a drug-carrying micelle can penetrate into a cell through the plasma membrane, degrade there, release the drug, and possibly escape to be recycled, as sketched in the left-hand panel of Fig. 8.9. By another mechanism, the micelle is incorporated into the plasma membrane built of the same lipids, leaving its cargo inside. This technique can also be used for diagnostic purposes and, most radically, for gene editing and reprogramming intracellular protein production by inserting genetic material in cells (Stewart et al, 2016). Intracellular delivery is still largely tested *in vitro* or in tissues extracted from an organism (*ex vivo*).

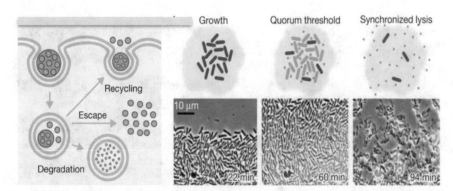

Fig. 8.9 *Left*: Intracellular delivery of a drug by micelles penetrating the plasma membrane by *endocytosis*. *Right*: Stages of growth and synchronized drug release by bacteria

---

[1] This property finds wider applications – more on this in the next section.

Fig. 8.10 *Left and center*: Configurations of a soft microrobot comprising a flagellum attached to a bilayer head. *Right*: A soft microrobot reshaping as it moves across a rough terrain

The role of artificial nanoparticles is in some way similar to bacteria infecting a tissue or viruses penetrating a cell, even though they are sent with better intensions. Jeff Hasty and coworkers (Din et al, 2016) undertook to employ bacteria to deliver drugs. In order to arrange periodic release of a drug, they took advantage of the quorum-sensing ability of bacteria (Sect. 7.4), which enables them to regulate gene expression in a way that depends on the population density. The researchers genetically induced bacteria to produce anti-tumour toxin and to commit suicide by breaking down their cellular membrane (*lysis*), releasing the load at the critical density threshold. There were always few defectors that remained alive. These kept multiplying until they reached the threshold again, which led to periodic bacterial growth, lysis, and drug delivery, as shown in the right-hand panels of Fig. 8.9. The method was even tested on a hapless mouse suffering from cancer (perhaps, induced by the experimentalists).

It was a lucky combination, with bacteria both producing and delivering the therapeutic load. If a drug is produced in the lab, a courier must be found that is able to bring it to the intended address. Prospective candidates for this role are artificial swimmers, already mentioned in Sect. 5.6, capable of traveling through blood vessels, or microrobots crawling like bacteria through interstices within tissues. Microbes with flagella were the original inspiration, with cargo propelled by an attached flagellum imitated by a chain of magnetic nanoparticles which undulate in an oscillating magnetic field (Dreyfus et al, 2005). Magnetic locomotion devices further developed into soft-bodied microrobots with reconfigurable body plans made of elastomer embedded with magnetic microparticles. A bilayer structure with actuated and passive layers enables bending and folding of the soft material, as in the left-hand panel of Fig. 8.10, which can be controlled by a magnetic field, and is also sensitive to ambient conditions (Huang et al, 2016). Soft-bodied "robots" can adjust to different liquid and solid terrains, switch between swimming, walking, crawling, and rolling locomotion (Fig. 8.10, right), and pick-up, carry, and release cargo (Hu et al, 2018).

8.5 Tinkering with Nature

A chance discovery by Jozef Schell and Marc van Montagu's group (Zaenen et al, 1974) that the *Agrobacterium tumefaciens* microbe induces cancer in plants by inserting detrimental code sequences into the cell's genome was redirected to beneficial use, adding desirable genes in a cheap and easy way, say, to make plants disease-resistant. This led to the development of an entire industry producing genetically modified organisms (GMO), referred to by its detractors as "Frankenfood", forbidden in some countries, although its ill effects are not evident. This opens both heavenly and abysmal prospects, from eliminating genetic diseases to producing designer babies to creating half-human hybrids to releasing deadly viruses into the environment. Of course, abhorrent applications of genetic engineering will be or already are legally forbidden, but this will not deter curious clandestine researchers, still less plain criminals.

The next step – perhaps from earthly to alien life – was the synthesis of unnatural DNA containing "letters" different from the four present in the natural genome. Floyd Romesberg and coworkers (Zhang et al, 2017) inserted a synthetic base pair into the genome of *E. coli*, the favorite microbe of researchers. The alien form was healthy and able to store the information carried by the six-letter alphabet, transcribing the unnatural nucleotides into the codons of messenger RNA and using them to produce proteins containing unconventional amino acids. It didn't stop here. Steve

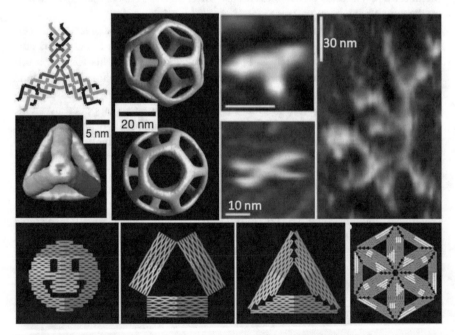

Fig. 8.11 *Top left*: A junction built from three single DNA strands and polyhedra built by their combinations. *Top right*: Branched DNA structures. *Bottom*: DNA origami

Benner's group (Hoshika et al, 2019) synthesized eight-letter DNA. What next? Romesberg (2019) declares: *Our goal is to further blur the boundary between the inanimate and the living, to create manmade parts that work within living systems, and thereby seamlessly alter or increase what they are capable of, ultimately creating new organisms with new attributes.* Nobody knows, of course, what the consequences will be. But once something becomes possible, it also becomes unstoppable.

A less dangerous direction is making use of genetic machinery for non-genetic aims. I mentioned in the preceding section how attached DNA strings can facilitate the formation of colloidal crystals. In this case, nanoparticles play a passive role: a variety of structures can be built from pure DNA, exploiting the ability of complementary nucleotide bases to selectively link together. Joining DNA strands by their sticky ends has a long history (Cohen et al, 1973); the conjunction was even inserted into *E. coli* and proved to be functional. In this way, it was also possible to assemble a cubic DNA molecule (Chen and Seeman, 1991) and "DNA origami" with fancy shapes (Rothemund, 2006), as shown in the bottom panels of Fig. 8.11. Far from being just toys, DNA origami were later proposed as vehicles for drug delivery (Jiang et al, 2012).

Other polyhedra were constructed (He et al, 2008) by preparing junctions built from three single strands, as shown in the upper left-hand panel of Fig. 8.11, to be placed at vertices of a three-dimensional supramolecule, like those shown on the same figure. Niles Pierce and coworkers (Yin et al, 2008) assembled the various structures shown in the upper right-hand panels of Fig. 8.11 through catalytic formation of branched junctions that spontaneously combine in irregular forms. They also synthesized an autonomous bipedal DNA walker capable of moving along a DNA track, just as natural protein molecular motors walk along protein filaments (Sect. 5.2). RNA structures can be designed and manipulated just as simply and flexibly, while retaining their catalytic functions (Guo, 2010).

Early diffraction images of microcrystalline DNA (Franklin and Gosling, 1953) were instrumental in the discovery of its double helix structure by James Watson and Francis Crick (1953). The design techniques developed in later years enabled self-assembly of varied crystalline structures (Paukstelis and Seeman, 2016). DNA crystals were proposed as scaffolds for orienting and positioning guest molecules, such as protein enzymes, and smart porous materials for molecular separation. These tasks are made precise thanks to the rigidity of DNA molecules and the specificity of interactions between base pairs. But the ultimate unnatural application of specific interactions between nucleotide bases is *molecular computing*.

Already at the dawn of computer technology, Richard Feynman (1961) envisaged miniaturization going right down to the molecular level. DNA is, in its essence, a digital coding element, and can be viewed therefore as a natural tool for molecular computing. Duplication or protein coding proceeds, like a Turing machine, by scanning DNA as a data tape. The first attempt to use DNA for computation was undertaken by Leonard Adleman (1994). He set out to solve the classical traveling salesman problem, that is, to find a route on a graph that visits each node exactly once, which is called a *Hamiltonian path*. The way to do this was first to synthesize

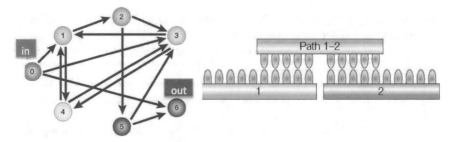

Fig. 8.12 *Left*: A wrong path through a seven-node graph. *Right*: Connections between Adleman's nucleotide sequences

seven random 20-base DNA strands representing each of seven nodes in a modest puzzle to be solved. Another set of 20-mers representing links contained ten nucleotides complementing the last ten of one node followed by another ten complementing the first ten of another node. When allowed to connect ("ligate") in accordance with complementation rules, as sketched in Fig. 8.12, these strands formed chains passing through the nodes – but not necessarily through all of them and not necessarily once. The path shown in the left-hand panel is *not* a Hamiltonian path. A lot of chemical operations followed: first to isolate chains containing exactly 140 bases, and then to eliminate those containing double strands and therefore passing some nodes twice. It took seven days of lab work to solve a problem that could be solved by hand much faster.

This could be ground-breaking work, however. Ruben and Landweber (2000) declared in their review: *We could be staring into the abyss of a science as doomed as phrenology or mesmerism. But we may be at the forefront of a new and creative technology whose implications have not even been fully mapped out, let alone realized.* They estimated that, if each ligation counts as a digital operation, the speed of the DNA computer far exceeded the speed of electronic computers of the time, and the energy efficiency, 2×10^{19} operations per joule, is only one order of magnitude less than the theoretical maximum set by the second law of thermodynamics, many orders of magnitude less energy than electronic computers waste. It was clumsy chemical post-processing that slowed down the results.

Where are we standing now, a quarter of a century later? It has been proven that DNA circuits can perform logic operations forming AND, OR, and NOT gates, can implement signal restoration, amplification, feedback, and cascading (Seelig et al, 2006), but molecular computation is far from being able to compete with electronic computers. Programming amounts to choosing appropriate double-stranded DNA, and an output molecule encodes the result; the same logical operation can often be carried out by assigning different nucleotide sequences. The "computation" is massively parallel, but the molecules are even more identical than worker ants, and even if everything runs fast and consumes very little energy (not counting what is used to run the lab), they all "solve" the same problem; the multitude just makes chemical analysis of the "result" possible.

Molecular computation might be efficient only if it is fed back into a biological environment. Amir et al (2014) used folded strands of DNA to create nanorobots which, depending on their interactions with proteins, change their conformation to release their payload. Differently configured robots emulate various logic gates. They don't need to be identical to ease chemical analysis, as in Adleman's experiment. On the contrary, their advantage is that they release different chemicals as the outcome of computations, and communicate with each other in this way. Like bacteria, these robots are autonomous and may self-organize their actions. The robots were injected into a live cockroach to deliver a molecule that targets its cells. The imagined perspective is to have a company of robots carrying a suite of drugs and deciding, on the basis of the local chemical environment, which ones to release – they would certainly know better than a doctor standing nearby and relying, at best, on the results of various tests averaged over the patient's whole body. They could even feel the patient's mental states, as in experiments by the same group sending electromagnetic signals to origami robots tethered to metal nanoparticles while circulating within a cockroach.

8.6 From Natural to Unnatural

Tinkering with Nature through the various modifications of the DNA code might be the most radical, most beneficial, and most dangerous perspective of 21st century technology. It is notable that it is driven by developing molecular *hardware* that at the same time carries software functions. This should be a sobering thought for those dealing with computer codes or words who are apt to confuse them with reality. Ray Kurzweil (2005) writes that *bits of information [. . .] is what we comprise*, and cites with gusto a line by the radical poet Muriel Rukeyser: *the universe is made of stories, not of atoms*. It is easy to forget that the explosive growth of computer power was due to anonymous creators of processors and memory elements rather than famous billionaire heads of information and software companies and *nouveau riche* start-uppers. Likewise, life evolved to higher forms by developing "hardware", the proteins forming cell machinery, and metabolic and communication networks. Software and finance, like the "selfish genes" promulgated by Dawkins (1976), are mere organizing structures parasitizing on material production and unable to function without its fruits.

Nanotechnology may have an even larger impact on explosive developments in the 21st century than artificial intelligence (AI). Everything really keeps accelerating, as if rushing into a singularity. Kurzweil, who was already predicting in 2005 that *the singularity is near* links it with AI, and compares it to the cosmic singularity – a black hole. This sounds quite grim if we recall the fate of matter falling into a black hole. Should we rejoice in the demise of our civilization at the singular point it is rushing into?

Kurzweil writes that *there will be no distinction, post-Singularity, between human and machine or between physical and virtual reality*. Youths immersed in computer games might be precursors of this Brave New World; with their way of life and the education they get they will need AI to care for them – but why should AI care about maintaining obsolete electrochemical appliances, much less providing them with *inexpensive vehicles with nanoengineered microwings* (which, as Kurzweil implies, should be able to carry people not yet miniaturized to microbes)? Why would it care for hungry uneducated masses? Would it maintain zoos housing creatures stubbornly clinging to their antiquated habits – Buddhist monks, Amish, orthodox Jews, recluses, remnants of hunter-gatherer tribes? *Homo Deus* by Yuval Noah Harari (2016), predicting the same kind of future, recounts a grim dystopia, whether intended or not.

The acolytes of exponential growth, while ridiculing linear forecasters, commit a similar error: they expect the exponent to remain constant, but it cannot. When any process appears to be rushing to the skies, something must happen on the way. In the best case, the exponential growth saturates, and something new and unpredictable emerges, and perhaps starts growing exponentially itself. In the worst case, the system explodes. History tells us that many prophecies have never materialized; *the future is hard to see*, as a simple-minded song reminds us, and still harder for professional prophets with their ingrained opinions and prejudices.

The image of human-like robots ubiquitous in science fiction is naive. The super-intellect of Solaris (Lem, 1961), mentioned in Sect. 7.4, may reside not in the structureless ocean but in a post-Singularity machine hidden underneath, either evolved from or created by its extinct biological precursors. On our planet and in our time, worldwide AI, at least in an embryonic form, already exists, hidden in Google's computer network and its rivals; it collects information about our movements and actions, and already surreptitiously governs the world. Self-driving cars, domestic robotic appliances, air navigation systems, you name it, are extensions of a remote AI. Their in-built processors, like the ant's brain, are too weak for their range of operation: for example, a car's computer couldn't hold the city map and traffic density. Like worker ants, they are parts of a remotely interacting assemblage, but, unlike ants, they do not self-organize, but are directed by the mighty omniscient remote processor. Only protected networks disconnected from internet for fear of hacking are out of reach.

AI still needs human help for maintenance, but takes over much of the programming. It also directs scientific research and development, serving its needs through a scientific communication network. A run-of-the-mill scientist is as minute a part of this network as an ant in its anthill. The paradigm-changing discoveries of the last century are receding into the past. There is a hope, at least, that scientists, like ants, self-organize – but do they? Or only in their private lives? The paths adopted by their foraging are laid out by funding agencies, and by university and government bureaucracies which, lacking independent judgement, are directed by the same remote processors collecting mountains of trivial statistics.

Biomorphic technology may eventually bring us brain-enhancing implants and rejuvenation treatment for the rich, while the masses failing to reach the *Homo Deus* status will, in the best (for AI) case, be drugged by electronic entertainment (as many already are), or in the worst case, rebel, bringing down the whole house of cards. A timely reaction, either by changed public attitudes or by governments already starting to restrain cyber-monopolists, may still redirect the nascent AI to more trivial and servile tasks. I don't mean to prophesy but my feeling is that in the end biology and electrochemistry, the material union of software and hardware in a self-contained individual, as a more resilient base of cognition, will win over one way or another, in the best case, by honing the molecular machinery of life, and in the worst, by rebooting evolution from an earlier stage.

Correction to: Broken Symmetry

Correction to:
Chapter 3 in: L. Pismen, *Morphogenesis Deconstructed*,
The Frontiers Collection,
https://doi.org/10.1007/978-3-030-36814-2_3

In the original version of the book, Fig. 3.5 has been replaced with revised figure in Chapter 3. The erratum chapter and the book have been updated with the change and approved by the author.

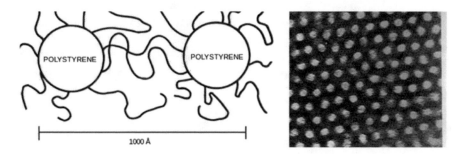

Fig. 3.5 *Left*: A scheme of separation of block copolymer units. *Right*: The resulting pattern

The updated version of this chapter can be found at
https://doi.org/10.1007/978-3-030-36814-2_3

References

Adleman LM, 1994. Science **266**, 1021–1024

Aigouy B, Farhadifar R, Staple DB, Sagner A, Roper JC, Jülicher F, 2010. Cell **142**, 773–786

Allen JJ, Bell GRR, Kuzirian AM, Hanlon RT, 2013. J. Morphology **274**, 645–656

Alon U, 2007. *An Introduction to Systems Biology: Design Principles of Biological Circuits*, Chapman & Hall, Boca Raton

Amir Y, Ben-Ishay E, Levner D, Ittah S, Abu-Horowitz A, Bachelet I, 2014. Nature Nanotechnology **9**, 353–357

Anderson LR, Owens TW, Naylor TJ, 2014. Biophys. Rev. **6**, 203–213

Anfinsen CB, 1973. Science **181**, 223–230

Arrhenius S, 1908. *Worlds in the Making: The Evolution of the Universe*, Harper & Row, New York

Arsenault AC, Puzzo DP, Manners I, Ozin GA, 2007. Nature Photonics **1**, 468–472

Ashlock D, 2006. *Evolutionary Computation for Modeling and Optimization*, Springer

Averbukh I, Ben-Zvi, D, Mishra S, Barkai N, 2014. Development **141**, 2150–2156

Avery OT, MacLeod CM, McCarty M, 1944. J. Experimental Medicine **79**, 137–158

Barrick JE, Yu DS, Yoon SH, Jeong H, Oh TK, Schneider D, Lenski RE, Kim JF, 2009. Nature **461**, 1243

Behe MJ, 2019. *Darwin Devolves*, HarperOne

Beloussov LV, Saveliev SV, Naumidi II, Novoselov VV, 1994. Int. Rev. Cytol. **150**, 1–34

Beloussov LV, 2015. *Morphomechanics of Development*, Springer

Bénard H, 1900. Ann. Chim. Phys. **7** (Ser. 23), 62

Ben-Jacob E, Shochet O, Tenenbaum A, Cohen I, Czirok A, Vicsek T, 1994. Nature **368**, 46–49

Ben-Jacob E, Cohen I, Gutnick DL, 1998. Annu. Rev. Microbiol. **52**, 779–806

Ben-Zvi D, Shilo BZ, Fainsod A, Barkai N, 2008. Nature **453**, 1205–1211

Ben-Zvi D and Barkai N, 2010. Proc. Nat. Acad. Sci. USA, **107**, 6924–6929

Bergert M, Erzberger A, Desai RA, Aspalter IM, Oates AC, Charras G, Salbreux G, Paluch EK, 2015. Nature Cell Biology **17**, 524–529

© Springer Nature Switzerland AG 2020

L. Pismen, *Morphogenesis Deconstructed*, The Frontiers Collection,

https://doi.org/10.1007/978-3-030-36814-2

Bernard C, 1879. *Leçons sur les phénomènes de la vie communs aux animaux et aux végétaux*, Librarie Baillière et Fils, Paris

Bernstein H, Byerly HC, Hopf FA, Michod RE, 1984. J. Theor. Biol. **110**, 323–351

Biot MA, 1941. J. Appl. Phys. **12**, 155–164

Bohr H, 1925. Acta Math. **45**, 580

Bonner JT, 2009. *The Social Amoebae: The Biology of Cellular Slime Molds*, Princeton University Press

Bratsun DA, Krasnyakov IV, Pismen LM, 2020. Biomech. Model. Mechanobiol., in print, DOI 10.1007/s10237-019-01244-z

Bravais A, 1850. J. École Polytech. **19**, 1–128

Brinkmann H, Baurain D, Philippe H, 2007. In *Planetary Systems and the Origins of Life*, eds. Pudritz RE, Stone JR, Higgs PG, Cambridge University Press

Browne EN, 1909. J. Exp. Zool. **7**, 1–23

Burgers JM, 1940. Proc. Phys. Soc. **52**, 23–33

Camacho-Lopez M, Finkelmann H, Palffy-Muhoray P, Shelly M, 2004. Nature Materials, **3**, 307–310

Canny MJ, 1977. Ann. Rev. Fluid Mech. **9**, 275–296

Čapek K, 1920. *R.U.R*, transl. Dover Publications, 2001

Capra F, 1975. *The Tao of Physics*, Shambhala, Boston

Chen J and Seeman NC, 1991. Nature **350**, 631–633

Chen Q, Bae SC, Granick S, 2011. Nature **469**, 381–384

Child CM, 1941. *Patterns and Problems of Development*, The University of Chicago Press

Choi S, Choi J, [...], Lukin MD, 2017. Nature **543**, 221–225

Chorost M, 2011. *World Wide Mind. The Coming Integration of Humanity, Machines, and the Internet*, Free Press, New York

Churchill FB, 2010. J. History Biol. **43**, 767–800

Cohen SN, Chang, ACY, Boyer HW, Helling RB, 1973. Proc. Natl. Acad. Sci. USA **70**, 3340–3344

Crick FH, 1970. Nature **225**, 420–422

Crick FH and Orgel LE, 1973. Icarus, **19**, 341–348

Damasio A, 2018. *The Strange Order of Things: Life, Feeling, and the Making of Cultures*, Pantheon, New York

Darwin C, 1859. *On the Origin of Species by Means of Natural Selection, or the Preservation of Favoured Races in the Struggle for Life*, John Murray, London

Davidson EH et al, 1995. Science **270**, 1319–1325

Dawkins R, 1976. *The Selfish Gene*, Oxford University Press

de Gennes PG, 1975. C. R. Acad. Sci. Ser. B **281**, 101–103

de la Loza MCD, Thompson BJ et al, 2017. Mechanisms of Development, **144**, 23–32

Diener TO, 1989. Proc. Natl. Acad. Sci. USA. **86**, 9370–9374

Din MO, Danino T, [...], Hasty J, 2016. Nature **536**, 81–85

Dixon HH, 1914. *Ascent of Sap in Plants*, Macmillan, London

Dixon RMW, 1997. *The Rise and Fall of Languages*, Cambridge University Press

Djokic T, Van Kranendonk MJ, Campbell KA, Walter MR, Ward CR, 2017. Nature Communications **8**, 15263

Dogterom M and Koenderink GH, 2019. Nature Reviews Molecular Cell Biology **20**, 39–54

Douady M and Couder Y, 1992. Phys. Rev. Lett. **68**, 2098–2101

Dreyfus R, Baudry J, Roper ML, Fermigier M, Stone HA, Bibette J, 2005. Nature **437**, 862–865

Dunn CW and Hejnol A, 2016. Current Biology **26**, R408–R431

Dyson FJ, 1982. J. Mol. Evol. **18**, 344–350

Dyson FJ, 2004. *Origins of Life*, Cambridge University Press

Eigen M, 1971. Naturwissenschaften **58**, 465–523

Eigen M, Gardiner W, Schuster P, Winkleroswatitsch R, 1981. Scientific American **244**, 88–92

Eigen M, 2002. Proc. Natl. Acad. Sci. USA **99**, 13374–13376

Eldredge N and Gould SJ, 1972. In *Models in Paleobiology*, TJM Schopf, ed., Freeman Cooper, San Francisco

Engelmann TW, 1883. Pflügers Arch. **30** (1), 95–124

Faraday M, 1831. Phil. Trans. Roy. Soc. London **121**, 299–318

Farhadifar R, Roper JC, Algouy B, Eaton S, Jülicher F, 2007. Curr. Biol. **17**, 2095–2104

Feynman RP, 1961. In *Minaturization*, pp 282–296, Gilbert DH (ed.), Reinhold, New York

FitzHugh R, 1961. Biophys. J. **1**, 445–466

Fletcher DA and Mullins D, 2010. Nature **463**, 485–492

Franklin RE and Gosling RG, 1953. Nature **171**, 740–741

Frohnhöfer HG and Nüsslein-Volhard C, 1986. Nature **324**, 120–125

Gardner M, 1970. Scientific American **223**, 120–123

Ge Q, Sakhaei AH, Lee H, Dunn CK, Fang NX, Dunn ML, 2016. Scientific Reports **6**, 31110

Geiger B, Spatz JP, Bershadsky AD, 2009. Nature Reviews Molecular Cell Biology **10**, 21–33

Gladman AS, Matsumoto EA, Nuzzo RG, Mahadevan L, Lewis JA, 2016. Nature Materials **15**, 413–418

Glendenning NK, 2001. Phys. Rep. **342**, 393–447

Golovin AA, Nepomnyashchy AA, and Pismen LM, 2002. Int. J. Bifurc. Chaos **12**, 2487–2500

Gordon DM, 1999. In *Information Processing in Social Insects*, Detrain C et al (eds.), Springer, Basel

Gould SJ, 1989. *Wonderful Life: The Burgess Shale and the Nature of History*, WW Norton, New York

Guo P, 2010. Nature Nanotechnology **5**, 833–841

Gutzeit HO, 1990. Eur. J. Cell Biol. **53**, 349–356

Haeckel E, 1874. *Anthropogenie oder Entwicklungsgeschichte des Menschen* (The Evolution of Man), Leipzig

Haeckel E, 1878. *Das Protistenreich*, Ernst Günther's Verlag, Leipzig

Hanczyc MM, Fujikawa SM, Szostak JW, 2003. Science **302**, 618–622

Harari YN, 2016. *Homo Deus. A Brief History of Tomorrow*, Harvill Secker

Harrington MJ, Razghandi K, Ditsch F, Guiducci L, Rueggeberg M, Dunlop JWC, Fratzl P, Neinhuis C, Burgert I, 2011. Nature Communucations **2**, 337

He B, Doubrovinski K, Polyakov O, Wieschaus E, 2014. Nature **508**, 392–396

He Y, Ye T, Su M, Zhang C, Ribbe AE, Jiang W, Mao C, 2008. Nature **452**, 198-201

Heidegger M, 1962. *Being and Time*, Harper, New York

Hershko A and Ciechanover A, 1998. Annu. Rev. Biochem. **67**, 425–479

Hessel JFC, 1831. *Krystallometrie, oder Krystallonomie und Krystallographie*, Leipzig

Hinton HE and Jarman GM, 1972. Nature **238**, 160–161

Hodgkin AL and Huxley AF, 1952. J. Physiol. London, **117**, 500–544

Hoffmann A and Tsonis PA, 2013, in *Pattern Formation In Morphogenesis*, eds Capasso V et al, pp 7–15

Hoffman BD, Grashoff C, Schwartz MA, 2011. Nature **475**, 316–323

Hoshika S, Leal NA, [. . .], Benner SA, 2019. Science **363**, 884–887

Howard J, 2001. *Mechanics of Motor Proteins and the Cytoskeleton*, Sinauer Associates, Sunderland Massachusets

Hu WQ, Lum GZ, Mastrangeli M, Sitti M, 2018. Nature **554**, 81–85

Huang HW, Sakar MS, Petruska AJ, Pane S, Nelson BJ, 2016. Nature Communications **7**, 12263

Ionov L, 2013. Adv. Funct. Mater. **23**, 4555–4570

Ising E, 1925. Beitrag zur Theorie des Ferromagnetismus, Z. Phys. **31** (1), 253–258

Jiang Q, Song C, [. . .], Ding B, 2012. J. Am. Chem. Soc. **134**, 13396–13403

Johnson S, 1755. *Dictionary of the English Language*, London

Kant I, 1755. *Allgemeine Naturgeschichte und Theorie des Himmels*, Königsberg, transl. http://records.viu.ca/ johnstoi/kant/kant2e.htm

Keating SJ, Gariboldi MI, Patrick WG, Sharma S, Kong DS, Oxman N, 2016. PLoS One **11**, e0160624

Keller L and Surette MG, 2006. Nature Rev. Microbiology **4**, 249

Keller RE, 1978. J. Morph. **157**, 223–248

Keren K, Pincus Z, Allen GM, Barnhart E, Marriott G, Mogilner A, Theriot JA, 2008. Nature **453**, 475–480

Kleman M, 1983. *Points, Lines and Walls*, Wiley, Chichester

Köpf MH and Pismen LM, 2013. Soft Matter **9**, 3727–3734

Kurzweil R, 2005. *The Singularity Is Near: When Humans Transcend Biology*, Viking Penguin

Lamarck JB, 1809. *Philosophie Zoologique*, Paris, trans. *The Zoological Philosophy*, Macmillan, London (1914)

Langer JS, 1980. Rev. Mod. Phys. **52**, 1–28

Laplace PS, 1796. *Exposition du Système du Monde*, Paris

Lauga E, 2016. Annu. Rev. Fluid Mech. **48**, 105–130

Lecuit T, 2013. in *Pattern Formation in Morphogenesis*, eds Capasso V et al, pp 41–57

Lee JM, 2013. *The Actin Cytoskeleton and the Regulation of Cell Migration*, Morgan & Claypool

Lee Y, Lee H, Hwang T, Lee JG, Cho M, 2015. Scientific Reports **5**, 16544

Lem S, 1961. *Solaris*, trans. 1970, Faber and Faber, London

Lendlein A, Jiang HY, Junger O, Langer R, 2005. Nature **434**, 879–882

Lent J, 2017. *The Patterning Instinct*, Blackstone Publishing, Ashland, Oregon

Leyser O and Domagalska M, 2011. Nature Reviews Molecular Cell Biology **12**, 211–221

Lighthill J, 1975. *Mathematical Biofluiddynamics*, SIAM, Philadelphia

Liu J, Prindle A, Humphries J, Gabalda-Sagarra M, Asally M, Lee DD, Ly S, Garcia-Ojalvo J, Süel GM, 2015. Nature **523**, 550–554

Lotka AJ, 1910. J. Phys. Chem. **14**, 271–274

Lovejoy AO, 1965. *The Great Chain of Being*, Harper and Row, New York

Lovelock J, 1979. *Gaia: A New Look at Life on Earth*, Oxford University Press

Lovelock J, 2007. *The Revenge of Gaia*, Penguin

Lowenberg C, Balk M, Wischke C, Behl M, Lendlein A, 2017. Acc. Chem. Res. **50**, 723–732

Lukin MD, 2017. Nature **543**, 221–225

Lwoff A, 1943. *L'Évolution Physiologique*, Paris

Mandelbrot BB, 1982. *The Fractal Geometry of Nature*, WH Freeman, San Francisco

Marchetti MC, Joanny JF, Ramaswamy S, Liverpool TB, Prost J, Rao M, Simha RA, 2013. Rev. Mod. Phys. **85**, 1143

Margulis L, 1970. *Origin of Eukaryotic Cells*, Yale University Press, New Haven

Margulis L, 1997. In Margulis L and Sagan D, 1997. *Slanted Truths*, Springer, New York

Margulis L, 1998. *Symbiotic Planet: A New Look at Evolution*, Basic Books, New York

Maynard Smith J and Szathmáry E, 1997. *The Major Transitions in Evolution*, Oxford University Press

McMenamin MAS, 2018. *Deep Time Analysis: A Coherent View of the History of Life*, Springer, Cham, Switzerland

Meinhardt H, 1982. *Models of Biological Pattern Formation*, Academic Press, London

Mereschkowski C, 1905. Biologisches Zentralblatt **25**, 593–604

Meron E, 2018. Annu. Rev. Cond. Mat. Phys. **9**, 79–103

Mertens F and Imbihl R, 1994. Nature **370**, 124–126

Milinkovitch MC et al, 2015. Nature Comm. **6**, 6368

Miller SL, 1953. Science **117**, 528–529

Mills DR, Peterson R, Spiegelman S, 1967. Proc. Natl. Acad. Sci. USA **58**, 217–224

Mogilner A and Oster G, 1996. Biophys. J. **71**, 3030–3045

Monier B, Gettings M, Gay G, Mangeat T, Schott S, Guarner A, Suzanne M, 2015. Nature **518**, 1–4

Moreno JL, 1934. *Who Shall Survive?*, Beacon House

Morgan TH, 1916. *A Critique of the Theory of Evolution*, Princeton University Press

Morozov KI and Leshansky AM, 2019. Langmuir **35**, 3987–3991

Newell AC, Shipman PD, Sun Z, 2008. J. Theor. Biology **251**, 421–439

Noor N, Shapira A, Edri R, Gal I, Wertheim L, Dvir T, 2019. Adv. Sci. **6**, 1900344

Northcutt RG, 2012. PNAS **109** (Suppl. 1) 10626–10633;
  https://doi.org/10.1073/pnas.1201889109

Nüsslein-Volhard C, 2006. *Coming to Life: How Genes Drive Development*, Kales Press, San Diego CA

Odell GM, Oster G, Alberch P, Burnside B, 1981. Delelopmental Biol. **85**, 446–462

Ohm C, Brehmer M, Zentel R et al, 2012. Adv. Polym. Sci. **250**, 49–94

Onsager L, 1944. A two-dimensional model with an order–disorder transition, Phys. Rev. **65**, 117

Oparin AI, 1924. *Proiskhozhdenie zhizni* (in Russian). Izd. Moskovskii Rabochii, Moscow. Transl. 1938, *The Origin of Life*, Macmillan, New York

Oster GF and Wilson EO, 1978. *Caste and Ecology in the Social Insects*, Princeton University Press

Ostwald W, 1900. Z. Phys. Chem. **35**, 33–76

Oxman N, 2017. Architectural Design **87**, SI 16–25

Palagi S, [. . .], Lauga E, Fischer P, 2016. Nature Materials **15**, 647–653

Paley W, 1809. *Natural Theology: or, Evidences of the Existence and Attributes of the Deity*, J Faulder, London

Pasteur L, 1864. *Conférences faites aux soirées scientifiques de la Sorbonne*

Paukstelis PJ and Seeman NC, 2016. Crystals **6**, 97

Pinheiro VB, Taylor AI, [. . .], Holliger P, 2012. Science **336**, 341–344

Pinker S, 1999. *Words and Rules: The Ingredients of Language*, Harvard University Press

Pismen LM and Simakov DSA, 2011. Phys. Rev. E **84**, 061917.

Pourquie O, 2003. Int. J. Dev. Biol. **47**, 597

Purcell EM, 1977. Am. J. Phys. **45**, 3–11

Qiu T, Lee TC, Mark AG, Morozov KI, Munster R, Mierka O, Turek S, Leshansky AM, Fischer P, 2014. Nature Communications **5**, 5119

Raab N, Davis J, Spokoini-Stern R, Kopel M, Banin E, Bachelet I, 2017. A symbiotic-like biologically-driven regenerating fabric, Scientific Reports **7**, 8528

Rayleigh Lord, 1916. On convection currents in horizontal layer of fluid when the higher temperature is on the under side. Philos. Mag. Ser. 6, **32** (192), 529–546

Remak R, 1855. *Untersuchungen über die Entwickelung der Wirbelthiere*, G Reimer, Berlin

Rich A, 1962. In *Horizons in Biochemistry*, pp 103–126, Academic Press, New York

Romesberg FE, 2019. Isr. J. Chem. **59**, 91–94

Rothemund PWK, 2006. Nature **440**, 297–302

Ruben AJ and Landweber LF, 2000. Nature Reviews Molecular Cell Biology **1**, 69–72

Ruiz-Herrero T, Fai TG, Mahadevan L, 2019. Phys. Rev. Lett. **123**, 038102

Russell B, 1972. *History of Western Philosophy*, Simon & Schuster

Saffman PG and Taylor G, 1958. Proc. Roy. Soc. A **245**, 312–329

Schelling TC, 1969. Amer. Econ. Rev. **59**, 488–493

Schoute JC, 1913. Rec. trav. bot. Neerl. **10**, 153–325

Schrödinger E, 1944. *What Is Life?*, Macmillan, Dublin

Schwarz US and Safran SA, 2013. Rev. Mod. Phys. **85**, 1327–1381

Seelig G, Soloveichik D, Zhang DY, Winfree E, 2006. Science **314**, 1585–1588

Shapiro JA, 1998. Annu. Rev. Microbiol. **52**, 81–104

Shechtman D, Blech I, Gratias D, Cahn JW, 1984. Phys. Rev. Lett. **53**, 1951–1953

Shipman PD, Sun Z, Pennybacker M, Newell AC, 2011. Eur. Phys. J. D **62**, 5–17

Simakov DSA, Cheung LS, Pismen LM, Shvartsman SY, 2012. Development **139**, 2814–2820

Sornette D, 2003. *Why Stock Markets Crash*, Princeton University Press

Sowa Y, Rowe AD, Leake MC, Yakushi T, Homma M, Ishijima A, Berry RM, 2005. Nature **437**, 916–919

Spemann H and Mangold H, 1924. Roux Arch. f. Entw. mech. **100**, 599–638

Stein A, Wilson BE, Rudisill SG, 2013. Chem. Soc. Rev. **42**, 2763–2803

Stenhouse D, 1974. *The Evolution of Intelligence – A General Theory and some of Its Implications*, Allen and Unwin, London

Stent GS, 1969. *The Coming of the Golden Age*, Natural History Press, Garden City, NY

Stewart MP, Sharei A, Ding X, Sahay G , Langer R, Jensen KF, 2016. Nature **538**, 183–192

Szostak JW and Zhu TF, 2009. J. Am. Chem. Soc. **131**, 5705–5713

Taber LA, 2009. Phil. Trans. Roy. Soc. A **367**, 3555–3583

Tallinen T, Chung JY, Rousseau, Girard N, Lefevre J, Mahadevan L, 2016. Nature Physics **12**, 588–593

Taylor AI, Pinheiro VB, Smola MJ, Morgunov AS, Peak-Chew S, Cozens C, Weeks KM, Herdewijn P, Holliger P, 2015. Nature **518**, 427–430

Tegmark M, 2014. *Our Mathematical Universe*, Alfred Knopf, New York

Tero A, Takagi S, Saigusa T, Ito K, Bebber DP, Fricker MD, Yumiki K, Kobayashi R, Nakagaki T (2010). Science **327**, 439

Thommes EW, 2008. In *Planetary Systems and the Origins of Life*, eds. Pudritz RE, Stone JR, Higgs PG, Cambridge University Press

Thomson W, 1871. On the equilibrium of vapour at a curved surface of liquid, Philosophical Magazine, series 4, **42** (282), 448–452

Thompson DW, 1917. *On Growth and Form*, Cambridge University Press

Timoshenko S, 1925. J. Opt. Soc. Am. **11**, 233–255

Trouilloud R, Yu TS, Hosoi AE, Lauga E, 2008. Phys. Rev. Lett. **101**, 048102

Trewavas A, 2014. *Plant Behavior and Intelligence*, Oxford University Press

Turing AM, 1952. Phil. Trans. Roy. Soc. London B **237**, 37–72

Tyree MT and Zimmermann MH, 2002. *Xylem Structure and the Ascent of Sap*, Springer, Berlin

Van Horn HM, 1968. Astrophys. J. **151**, 227–238

Vasquez CG and Martin AC, 2015. Nature **518**, 171–173

Vicsek T, Czirok A, Ben-Jacob E, Cohen I, Shochet O, 1995. Phys. Rev. Lett. **75**, 1226–1229

Viktorinova I, Pismen LM, Aigouy B, Dahmann C, 2011. J. R. Soc. Interface **8**, 1059–1063

Vilfan A, 2012. Eur. Phys. J. E **35**, 72

von Neumann J, 1951. *The general and logical theory of automata*. In LA Jeffress (Ed.), Cerebral mechanisms in behavior; the Hixon Symposium, pp. 1–41, Wiley, Oxford, England

von Nägeli C, 1884. *Mechanisch-physiologische Theorie der Abstammungslehre*, Verlag Oldenbourg, München

Wächtershäuser G, 1988. Microbiol. Rev. **52**, 452–484

Wächtershäuser G, 2007. Chemistry & Biodiversity **4**, 584–602

Waddington CH, 1942. Nature **150**, 563–565

Waddington CH, 1961. *The Nature of Life*, Unwin Books, London

Waters CM and Bassler BL, 2005. Annu. Rev. Cell Dev. Biol. **21**, 319–346

Watson JD and Crick FH, 1953. Nature **171**, 737–738

Weismann A, 1885. *The Continuity of the Germ-Plasm as the Foundation of a Theory of Heredity*, transl. 1891, in *Essays upon Heredity and Kindred Biological Problems*, Clarendon Press, Oxford, pp 163–256

Wilczek F, 2012. Phys. Rev. Lett. **109**, 160401

Williamson DI, 1992. *Larvae and Evolution: Toward a New Zoology*, Chapman and Hall, New York

Wilson EO, 2012. *The Social Conquest of Earth*, Liveright, New York

Wittgenstein L, 1922 *Tractatus Logico-Philosophicus*, Routledge and Kegan, London

Wolfram S, 2002. *A New Kind of Science*, Wolfram Media

Wolpert L, 1969. J. Theor. Biol. **25**, 1–47

Wolpert L, 2016. Current Topics in Developmental Biology **117**, 597–608

Xie T, 2010. Nature **464**, 267–270

Yin P, Choi HMT, Calvert CR, Pierce, NA, 2008. Nature **451**, 318–322

Yu CJ, Li YH, […], Rogers JA, 2014. Proc. Natl. Acad. Sci. USA **111**, 12998–13003

Zaenen I, van Larebeke N, Teuchy H, van Montagu M, Schell J, 1974. J. Molec. Biol. **86**, 109–116

Zhabotinsky AM, 1964. Biofizika **9**, 306–311

Zhang YK, Lamb BM, Feldman AW, Zhou AX, Lavergne T, Li LJ, Romesberg FE, 2017. Proc. Natl. Acad. Sci. USA **114**, 1317–1322

Zigmond S, 2004. Nature Cell Biology **6**, 12–14

Zöttl A and Stark H, 2018. Eur. Phys. J. E **41**, 61

Zwicker D, Seyboldt R, Weber CA, Hyman AA, Jülicher F, 2017. Nature Physics **13**, 408–413

Illustration Credits

Illustrations are published under the Creative Commons license, except for those where Springer holds the copyright, those produced by the author, and those where consent has been given by a copyright owner.

Figure 1.1(l): Public Domain, https://en.wikipedia.org/wiki/History_of_the_alphabet#/media/File:Ph%C3%B6nizisch-5Sprachen.svg
Figure 1.1(r): Public Domain, https://en.wikipedia.org/wiki/Genetic_code#/media/File:GeneticCode21-version-2.svg
Figure 2.1: By ESA/Hubble & NASA – http://www.spacetelescope.org/images/potw1108a/, Public Domain, https://commons.wikimedia.org/w/index.php?curid=13586042
Figure 2.2: By R.N. Bailey – Own work, CC BY 4.0,
https://commons.wikimedia.org/w/index.php?curid=59672008
Figure 2.3(l): By ALMA, CC BY 4.0, https://commons.wikimedia.org/w/index.php?curid=36643860
Figure 2.3(r): Public Domain, https://commons.wikimedia.org/wiki/File:A_Distant_Planetary_System.jpg
Figure 2.4(l): By Williams12357 – Own work,
CC BY-SA 3.0, https://commons.wikimedia.org/w/index.php?curid=19031519
Figure 2.4(r): Public Domain, https://commons.wikimedia.org/w/index.php?curid=3473857
Figure 2.6(l): Public Domain, https://commons.wikimedia.org/w/index.php?curid=5313203
Figure 2.5(r): By SergioJimenez – Own work, CC BY-SA 4.0,
https://commons.wikimedia.org/w/index.php?curid=45516736
Figure 2.7(r): By (MTheiler) – Own work, screen capture, CC BY-SA 4.0,
https://commons.wikimedia.org/w/index.php?curid=74697544
Figure 3.2(l,c): By Diamond_and_graphite.jpg: User:Itubderivative work: Materialscientist (talk) – Diamond_and_graphite.jpgFile:Graphite-tn19a.jpg, CC BY-SA 3.0,
https://commons.wikimedia.org/w/index.php?curid=7223557
Figure 3.2(r): By User Qwerter at Czech wikipedia: Qwerter. Transferred from cs.wikipedia to Commons by sevela.p. Translated to english by Michal Ma?as (User:snek01). Vectorized by Magasjukur2 – File:3D model hydrogen bonds in water.jpg, CC BY-SA 3.0,
https://commons.wikimedia.org/w/index.php?curid=14929959
Figure 3.3(l): By Inductiveload – Own work, Public Domain,
https://commons.wikimedia.org/w/index.php?curid=5839079
Figure 3.3(c): By Materialscientist – Own work, CC BY-SA 3.0,
https://commons.wikimedia.org/w/index.php?curid=10094837
Figure 3.4(l): By Psiñedelisto, based on version by Dbuckingham42 – Own work, CC BY-SA 4.0,
https://commons.wikimedia.org/w/index.php?curid=41724357

© Springer Nature Switzerland AG 2020
L. Pismen, *Morphogenesis Deconstructed*, The Frontiers Collection,
https://doi.org/10.1007/978-3-030-36814-2

Figure 3.4(c): By Wilson Bentley, From Annual Summary of the "Monthly Weather Review" for 1902., Public Domain, https://commons.wikimedia.org/w/index.php?curid=22130

Figure 3.4(r): By AMES lab., http://cmp.ameslab.gov/personnel/canfield/photos.html#, Public Domain

Figure 3.5(l): Public Domain, https://commons.wikimedia.org/w/index.php?curid=20540437

Figure 3.5(r): By Peter R Lewis, Public Domain, https://commons.wikimedia.org/w/index.php?curid=41030241

Figure 3.6(l): Adapted from Milinkovitch et al, 2015

Figure 3.6(r): Adapted from Hinton and Jarman, 1972

Figure 3.7: By Kebes – Own work, CC BY-SA 3.0, https://commons.wikimedia.org/w/index.php?curid=4170779, https://commons.wikimedia.org/w/index.php?curid=4170810, https://commons.wikimedia.org/w/index.php?curid=4170873

Figure 3.8(l): Courtesy Prof. Masao Doi

Figure 3.9(l): From Ohm et al, 2012

Figure 3.9(tr): From Palagi et al, 2016

Figure 3.9(br): From Camacho-Lopez et al, 2004

Figure 3.11(c,r): From Golovin et al, 2002. Int. J. Bifurc. Chaos, **12**, 2487–2500

Figure 3.12 By Javier Bartolomé Vílchez (Autor de la imagen original: David Gabriel García Andrade) – Modificación de una imagen subida al Commons, CC BY-SA 3.0, https://commons.wikimedia.org/w/index.php?curid=4450178

Figure 3.13(l): By Simpsons, Public Domain, https://commons.wikimedia.org/w/index.php?curid=18284509

Figure 3.13(r): From Mertens and Imbihl, 1994

Figure 3.15(l): By Paul from Enschede, The Netherlands – Dendritic Copper Crystals, CC BY-SA 2.0, https://commons.wikimedia.org/w/index.php?curid=18829701

Figure 3.15(c): Adapted from Tarafdar S et al, 2008. Eur. Phys. J. E **25**, 267-275

Figure 3.15(r): From Tallinen et al, 2016 https://commons.wikimedia.org/w/index.php?curid=205440

Figure 4.1: Public Domain, https://commons.wikimedia.org/w/index.php?curid=1201601

Figure 4.2(l): By File:Difference DNA RNA-DE.svg: Sponk / *translation: Sponk – Chemical structures of nucleobases by Roland1952, CC BY-SA 3.0, https://commons.wikimedia.org/w/index.php?curid=9810855

Figure 4.2(r): By Vossman – Own work, CC BY-SA 3.0, https://commons.wikimedia.org/w/index.php?curid=7115139

Figure 4.3(r): By Zephyris at the English language Wikipedia, CC BY-SA 3.0, https://commons.wikimedia.org/w/index.php?curid=2426907

Figure 4.4: By Andrew Z. Colvin – Barth F. Smets, Ph.D., with permission, CC BY-SA 4.0, https://commons.wikimedia.org/w/index.php?curid=69411999

Figure 4.5: By Oxygenation-atm.svg: Heinrich D. Hollandderivative work: Loudubewe (talk) – Oxygenation-atm.svg, CC BY-SA 3.0, https://commons.wikimedia.org/w/index.php?curid=12776502

Figure 4.5 (inset): By Doc. RNDr. Josef Reischig, CSc. – Author's archive, CC BY-SA 3.0, https://commons.wikimedia.org/w/index.php?curid=31550579

Figure 4.6: By SuperManu – Own work, CC BY-SA 3.0, https://commons.wikimedia.org/w/index.php?curid=2902736

Figure 4.7: From Zwicker et al, 2017

Figure 4.8: Public Domain, https://commons.wikimedia.org/w/index.php?curid=2145991

Figure 4.9: By Unknown – Originally published in The Hornet magazine; Public Domain, https://commons.wikimedia.org/w/index.php?curid=23436

Figure 4.10(l): By Didacus Valades (Diego Valades) – Rhetorica Christiana, Public Domain, https://commons.wikimedia.org/w/index.php?curid=1688250

Figure 4.10(r): By G Avery – Scientific American, Public Domain, https://commons.wikimedia.org/w/index.php?curid=18535798

Figure 5.1(l): Public Domain, https://commons.wikimedia.org/w/index.php?curid=2109671

Figure 5.1(r): By Martin Grandjean – Own work, CC BY-SA 4.0, https://commons.wikimedia.org/w/index.php?curid=39967993

Figure 5.2(l): By Louisa Howard, Public Domain, https://commons.wikimedia.org/w/index.php?curid=1246305

Figure 5.2(r): By Nucleus ER golgi.jpg: Magnus ManskeDerivative work: Pbroks13 (talk) – File:Nucleus ER golgi.jpg, CC BY 3.0, https://commons.wikimedia.org/w/index.php?curid=6208993

Figure 5.3(r): By Nikolya188811 – Own work, CC BY-SA 4.0, https://commons.wikimedia.org/w/index.php?curid=76098276

Figure 5.4: By OpenStax – https://cnx.org/contents/FPtK1zmh@8.25:fEI3C8Ot@10/Preface, CC BY 4.0, https://commons.wikimedia.org/w/index.php?curid=30131170

Figure 5.6(l): By Mariana Ruiz Villarreal LadyofHats, Public Domain, https://commons.wikimedia.org/w/index.php?curid=6195050

Figure 5.6(r): By Synaptidude, CC BY 3.0, https://commons.wikimedia.org/w/index.php?curid=21460910

Figure 5.7: Adapted from Original by en:User:Chris 73, CC BY-SA 3.0, https://commons.wikimedia.org/w/index.php?curid=44114666 and BruceBlaus, Own work, CC BY 3.0, https://commons.wikimedia.org/w/index.php?curid=29452220

Figure 5.8(l): By Thomas Splettstoesser (www.scistyle.com) – Own work (rendered with Maxon Cinema 4D), CC BY-SA 4.0, https://commons.wikimedia.org/w/index.php?curid=41014850

Figure 5.8(r): By http://rsb.info.nih.gov/ij/images/, Public Domain, https://commons.wikimedia.org/w/index.php?curid=655748

Figure 5.9(l): http://www.proteinatlas.org/dictionary/cell/focal+adhesions

Figure 5.9(rt): Goley ED and Welch MD, Nature Reviews Molecular Cell Biology, 7, 713-726 (2006)

Figure 5.9(rb): By Rpgch [CC BY-SA 4.0 (https://creativecommons.org/licenses/by-sa/4.0)], from Wikimedia Commons

Figure 5.10(l): By Hotulainen, Pirta, and Pekka Lappalainen. – "Stress fibers are generated by two distinct actin assembly mechanisms in motile cells." The Journal of cell biology 173.3 (2006): 383-394., CC BY-SA 3.0, https://commons.wikimedia.org/w/index.php?curid=38850981

Figure 5.10(r): By OpenStax – https://cnx.org/contents/FPtK1zmh@8.25:fEI3C8Ot@10/Preface, CC BY 4.0, https://commons.wikimedia.org/w/index.php?curid=30015035

Figure 5.11(l): From Fletche and Mullins, 2010

Figure 5.12: From Dogterom and Koenderink, 2019

Figure 5.13: Adapted from Ali Zifan – Own work, https://commons.wikimedia.org/w/index.php?curid=49721676

Figure 5.14(l): By Unknown, Public Domain, https://commons.wikimedia.org/w/index.php?curid=392819

Figure 5.14(r): By Rdbickel – Own work, CC BY-SA 4.0, https://commons.wikimedia.org/w/index.php?curid=49599354

Figure 5.15: From Anderson et al, 2014. Biophys. Rev. 6 203-213

Figure 5.16(tl): From Ben-Zvi et al, 2008

Figure 5.16(tr,b): From Bergert et al, 2015

Figure 5.17(l): Adapted from https://commons.wikimedia.org/w/index.php?curid=67552518

Figure 5.17(r): Adapted from Qiu et al, 2014

Figure 5.18(l): From Sowa et al, 2005

Figure 5.18(r): Adapted from Vilfan, 2012

Figure 5.19: From Zöttl and Stark, 2018

Figure 6.2: Katelynp1 – Own work, CC BY-SA 3.0, https://commons.wikimedia.org/w/index.php?curid=19543202

Figure 6.5: Lithograph by J. G. Bach of Leipzig after Haeckel (1874), Public Domain, https://commons.wikimedia.org/w/index.php?curid=8007834

Figure 6.3: Photos by Mark McMenamin, 2018

Figure 6.4(l): CC BY-SA 3.0, https://commons.wikimedia.org/w/index.php?curid=2173118
Figure 6.4(r): By Ian Alexander – Own work, CC BY-SA 4.0,
https://commons.wikimedia.org/w/index.php?curid=67051327
Figure 6.6: Public Domain, https://commons.wikimedia.org/w/index.php?curid=42441233
Figure 6.7: Adapted from Hoffmann and Tsonis, 2013
Figure 6.10: From Lecuit, 2013
Figure 6.12(t): Adapted from He et al, 2014
Figure 6.12(bl): Adapted from Vasquez and Martin, 2015
Figure 6.12(br): Adapted from Monier et al, 2015
Figure 6.14(l): Adapted from Geiger et al, 2009
Figure 6.14(r): From Zigmond, 2004
Figure 6.15: Adapted from Hoffman et al, 2011
Figure 6.16(l): CC BY-SA 3.0, https://commons.wikimedia.org/w/index.php?curid=121444
Figure 6.16(c): From Shipman et al, 2011
Figure 6.16(r): By Stan Shebs, CC BY-SA 3.0,
https://commons.wikimedia.org/w/index.php?curid=693560
Figure 6.18(l): From Leyser and Domagalska, 2011
Figure 6.18(r): Adapted from LadyofHats, Public Domain,
https://commons.wikimedia.org/w/index.php?curid=1685428
Figure 6.19: Adapted from Harrington et al, 2011
Figure 7.1(l): Public Domain, https://commons.wikimedia.org/w/index.php?curid=1215127
Figure 7.1(r): Adapted from Tyree and Zimmermann, 2002
Figure 7.2: Adapted from File:Blood vessels-en.svg: Kelvinsongderivative work: Begoon – This
file was derived from: Blood vessels-en.svg:, CC BY-SA 3.0,
https://commons.wikimedia.org/w/index.php?curid=65695195
Figure 7.3: By Egm4313.s12 (Prof. Loc Vu-Quoc) – Own work, CC BY-SA 4.0,
https://commons.wikimedia.org/w/index.php?curid=72816083
Figure 7.4: Adapted from Ben-Jacob et al, 1994
Figure 7.5: From Keller and Surette, 2006
Figure 7.6: From Liu et al, 2015
Figure 8.1(c): Public Domain, https://commons.wikimedia.org/w/index.php?curid=8873104
Figure 8.1(r): By Rama – Own work, CC BY-SA 2.0 fr,
https://commons.wikimedia.org/w/index.php?curid=2839956
Figure 8.2(l): Adapted from Lee et al, 2015
Figure 8.2(r): Adapted from Lendlein et al, 2005
Figure 8.3(l): By Neri Oxman CC BY-SA 4.0,
https://commons.wikimedia.org/w/index.php?curid=49856306
Figure 8.3(r): From Raab et al, 2017
Figure 8.4: From Ge et al, 2016
Figure 8.5: From Gladman et al, 2016
Figure 8.9(l): From Stewart et al, 2016
Figure 8.9(r): From Din et al, 2016
Figure 8.10(l,c): From Huang et al, 2016
Figure 8.10(r): From Hu et al, 2018
Figure 8.11(tl): From He et al, 2008
Figure 8.11(tr): From Yin et al, 2008
Figure 8.11(b): From Rothemund, 2006
Figure 8.12: From Ruben and Landweber, 2000

Printed in the United States
By Bookmasters